Economic Development of Socialist Vietnam, 1955-80

G. Nguyen Tien Hung

Economic Development of Socialist Vietnam, 1955-80

PRAEGER SPECIAL STUDIES IN INTERNATIONAL ECONOMICS AND DEVELOPMENT

Praeger Publishers New York London

Library of Congress Cataloging in Publication Data

G. Nguyen Tien Hung, 1935-
Economic development of socialist Vietnam, 1955-80.

(Praeger special studies in international economics and development)
Includes bibliographical references.
1. Vietnam--Economic conditions. 2. Vietnam--Economic policy. 3. Agriculture--Vietnam. 4. Vietnam--Industries. I. Title.
HC443.V5N47 1977 330.9'59704 77-11149
ISBN 0-275-24080-0

PRAEGER SPECIAL STUDIES
200 Park Avenue, New York, N.Y., 10017, U.S.A.

Published in the United States of America in 1977
by Praeger Publishers,
A Division of Holt, Rinehart and Winston, CBS, Inc.

789 038 987654321

Printed in the United States of America

To

My Parents

and

G. Warren Nutter

ACKNOWLEDGMENTS

No acknowledgment can give full credit to all those who have assisted me and left an imprint on my thinking.

I am grateful to the Earhart Foundation for the generous financial assistance that made the research on this project possible. I am particularly indebted to Richard Ware, president of the foundation, and to Professor G. Warren Nutter for their continuous encouragement and support, without which I might have had to abandon the project. This is particularly true when the data problem became overwhelming.

Personal thanks are due to Dr. Lorraine A. Williams for her keen interest in my research. This book is made possible with partial support from the Faculty Research Program in Social Sciences, Humanities, and Education at Howard University and from the Office of the Vice President for Academic Affairs.

I owe a debt of gratitude to my former colleague, the late Bernard Fall, for stimulating discussion during 1965-66 on the economic problems of North Vietnam. To the Department of Economics at Howard University, under the Ford Foundation program, my special thanks for providing me research and secretarial services for the completion of the manuscript. In this phase, I also benefited enormously from the research facility at the Thomas Jefferson Center Foundation. Comparative studies on other areas of social economics available at the foundation helped me to understand the relevant socialist statistical concepts.

In the initial stage of research, I was privileged to discuss some important aspects of the subject with several outstanding Vietnamese scholars, among them Nguyen Ngoc Bich, Hoang Van Chi, and Nguyen Manh Hung. I am grateful for their helpful comments, which enhance my understanding of the socialist transformation of Vietnam. My appreciation is also extended to Professor and Mrs. Lien-fu Huang for assisting me in statistical research.

I cannot express adequately how much I profited from my wife's assistance at every stage of research and finalizing the manuscript.

G. Nguyen Tien Hung
Howard University
Washington, D. C.

CONTENTS

LIST OF TABLES AND FIGURES

Table		Page

INTRODUCTION

Now that the unification of Vietnam is well underway, how is North Vietnam going to apply its development model to South Vietnam? What would be the impact of a unified Vietnamese economy on the Southeast Asian scene? On the Pacific area's development and stability? Would the former Indochinese Economic Union be reestablished, and if so, what is the implication for the rest of Asia? These are among the questions that will be of interest to students of Asian economic development and international development during the rest of the 1970s and into the 1980s.

Inquiries into these questions require as prerequisite an understanding of the type of economic system into which North Vietnam has been transformed, the pattern of economic development the country has undergone, and above all, the strengths and weaknesses of its economy. Over the past decade, political and military questions related to the war, and the war itself, dominated public debate and scholarly research on Vietnam. There has been very little study on the North Vietnamese economy, and rarely does it exist in English. The very limited availability of studies on the subject can also be explained by the lack of statistical data on North Vietnam, owing to the absolutely secretive nature of its regime, secretive even by Communist standards.

In this book, I attempt to undertake a first systematic study on the North Vietnamese economy based on materials appearing in Vietnamese, French, and, to a much lesser extent, English. Starting with data on the pre-Communist era as a base for comparison, painstaking effort has been made to collect and to analyze fragmentary data, to reconstruct figures from different official sources, and wherever possible to make inference in order to arrive at meaningful and comprehensive sets of time series, no matter how preliminary and incomplete, to serve as a basis for analysis.

The research, which I began in 1970, was motivated by my conviction that there was a very important economic factor underlying the Northern war effort: the traditional dependence of the North on the South to compensate for its annual food deficit, combined with the failure of the Communist regime to solve the food problem and its inability, in negotiating with the Southern government, to resume intraregional trade after the partition of the country in 1954.* Hence,

*See Appendix to Chapter 2 of this book.

total war was undertaken to conquer the South. This point will be emphasized by the data in Tables 1A. 1 and 2A. 1.

North Vietnam's total mobilization of manpower and resources during 1965-75 to sustain the war was finally rewarded with complete victory in the spring of 1975. It is noteworthy that the Northern commander who led the occupation of Saigon, General Van Tien Dung, is the same one who, twenty years earlier, had led the Northern delegation to the Southern capital for an abortive negotiation to resume economic trade between the two Vietnams.

Events since April 1975, together with the speed with which the North-South railroad (the main artery of interzonal trade) was rebuilt during 1976, have served to dramatize my thesis concerning the economic factor behind the war.

A thorough study of the Northern economy must begin with an examination of its relationship with the rest of Vietnam, that is, in the historical context of economic development prior to the partition in 1954.

The complementarity of the two regions, discussed in Chapter 1, together with the traditional flow of interregional trade, analyzed in Chapter 2, show the salient features of the initial conditions of the North Vietnamese economy.

The consequences of partition and the economic reconstruction that followed are considered in Chapter 3, and the paramount task of socialist transformation following the Sino-Soviet model is discussed in Chapter 4.

Economic-development planning toward unbalanced growth favoring heavy industry, which started in 1958 with a three-year plan, followed by a five-year plan, is examined in Chapter 5.

Overall economic performance during the past two decades is evaluated beginning with an estimate of trends in national income (Chapter 6), and is followed by an analysis of the two most important sectors: agriculture (Chapter 7) and industry (Chapter 8). Agriculture is emphasized because it has been and remains the most important sector in terms of income generation and employment, and, above all, as a way of life for the majority of the people.

Conquest of the South opens a new chapter in the economic development of the North and of the country as a whole. Unification brings with it prospects and problems that are analyzed in Chapter 9, which also discusses the application of the Northern model to the socialist transformation of the South.

Toward the future, the regime's grand design to industrialize the economy in order to build socialism is appraised through an analysis of the 1976-80 five-year plan.

Limitation of time precludes full elaboration of several important topics that are only touched upon here, such as economic efficiency (optimality), fiscal performance, and foreign trade and aid. These topics, I hope can be explored at a later date.

Vietnam

Source: U.S., Senate, Committee on the Judiciary, Subcommittee to Investigate Problems Connected with Refugees and Escapees, Aftermath of War: Humanitarian Problems of Southeast Asia, 94th Cong., 2d sess., May 17, 1976, p.12.

PART I

THE INITIAL CONDITIONS: VIETNAM DIVIDED

CHAPTER

1

THE TWO VIETNAMS

Although the economy of all Vietnam is largely agricultural, with that sector producing in prewar times some 65 percent of gross national product (GNP) and providing employment for over 90 percent of the population,[1] the country is clearly divided into two complementary regions—the mining North and the agricultural South.

The North, with 63,344 square miles, possesses nearly all of the country's mineral resources, while the South, with a surface area about 5 percent larger, 66,200 square miles, is abundant in cultivable land.

The structural differences in the two regions are illustrated in Tables 1.1 and 1.2. In Table 1.1, data for 1942 are used to illustrate the economic picture under normal conditions of peacetime; 1942 is also the last year for which data for all Vietnam are available. Estimates for 1955 show conditions roughly at the time of partition, while data for 1960 compare the two zones five years after the partition and represent the situation on the eve of North Vietnam's First Five-Year Plan. The 1965 figures represent the last year of normal performance in the economies of both zones during the postpartition period, for as the internal conflict intensified in 1966, agricultural production was sharply reduced. Because of the overwhelming importance of the role of agricultural production in the economy of all Vietnam, this chapter focuses mainly on the agricultural differences between the two Vietnams.

ECONOMIC COMPLEMENTARITY

While analysis of complementarity in this section focuses mainly on the initial condition—that is, the situation—as it existed prior to the partition of the country in 1954, it should be pointed out

TABLE 1.1

Distribution of Population, Area under Cultivation, and Production of Major Commodities in North and South Vietnam in Selected Years (area in thousand hectares; production in thousand metric tons)

Population Cultivated Area/Commodity	Quantity in North	Percent of Total	Quantity in South	Percent of Total	Total
		1942			
Population (thousands of persons)	12,550[a]	59	8,900[b]	41	21,450
Cultivated area	1,925	41	2,811	59	4,736
Paddy	2,335	39	3,711[e]	61	6,046
Rubber	870	0	51,760	100	52,630
Coal	2,329	100	0	0	2,329
Cement	270	100	0	0	270
		1955			
Population (thousands of persons)	13,574	53	11,948[c]	47	25,522
Cultivated area	1,600	42	2,179	58	3,779
Paddy	2,080[f]	43	2,767	57	4,847
Rubber	0	0	66,336	100	66,336
Coal	642	100	0	0	642
Cement	9	100	0	0	9
		1960			
Population (thousands of persons)	15,516	53	14,052	47	29,963
Cultivated area	2,036	45	2,471	55	4,507
Paddy	3,054[f]	36	5,421	64	8,475
Rubber	0	0	78,011	100	78,011
Coal	2,450	99	27	1	2,477
Cement	380	100	0	0	380
		1965			
Population (thousands of persons)	18,904[c]	54	15,900	46	34,804
Cultivated area	2,225	48	2,429	52	4,654
Paddy	3,338[f]	41	4,822	59	8,160
Rubber	0	0	60,900	100	60,900
Coal	4,000[h]	100	0	0	4,000
Cement	710	79	189	21	899

			1970 (estimated)		
Population (thousands of persons)	22,452[c]	56	17,333	44	39,785
Cultivated area	1,700[f]	41	2,430	59	4,130
Paddy	2,890[f]	36	5,115	64	8,005
Rubber	0	0	27,650	100	27,650
Coal	3,000	100	0	0	3,000
Cement	600[i]	68	286	32	886
			1973 (estimated)		
Population (thousands of persons)	24,893[c]	57	18,720[d]	43	43,613
Cultivated area	2,200[f]	39	3,500[g]	61	5,700
Paddy	3,960[f]	35	7,350[g]	65	11,310
Rubber	—	—	—	—	—
Coal	5,000[i]	100	0	0	5,000
Cement	750[i]	65	400[j]	35	1,150

[a]Of which Tonkin accounted for 9,920,000.

[b]Of which Cochinchina accounted for 5,400,000.

[c]Estimates based on official figures of previous years, especially the 1960 census, and the rate of population growth of 3.5 percent per annum.

[d]Estimated population based on growth rate of 2.6 percent per annum.

[e]Of which Cochinchina alone accounted for 3,179,000.

[f]Estimated by author; these figures are different from official data in Table 7.5.

[g]Estimates based on assumption that the current level of security is maintained or improved, and on current yield of 2.1 tons per hectare.

[h]Planned target was 5 million.

[i]Estimates based on North Vietnam's target of restoring the 1965 level by 1972-73.

[j]Rough projections.

Note: North and South are defined as in the 1954 Geneva agreement; North Vietnam includes Bac Bo (Tonkin) and four Northern provinces of TrungBo (Annam), and South Vietnam includes Nam Bo (Cochinchina) and Trung Bo (Annam) from the 17th parallel.

Sources: France, Haut Commissariat de France pour l'Indochine, Annuaire Statistique de i'Indochine, vol. 10, 1941-42, pp. 87-88, and vol. 11, 1943-46, pp. 271-95; Democratic Republic of Vietnam, (hereafter cited as DRVN), Central Statistical Office, Nam Nam Xay Dung Kinh Te va Van Hoa (Five years of economic and cultural building) (Hanoi: Central Statistical Office, 1960), pp. 71, 113-28, 163; Nghien Cuu Kinh Te (Economic Studies) (Hanoi) various issues, 1965-70; DRVN, So Lieu Thong Ke (Statistical data) (Hanoi, 1963); Le Monde, December 24 and 25, 1970, p. 4; Tiers Monde 11, no. 42-43 (April-September, 1970): 532-34; U.S., Agency for International Development, Annual Statistical Bulletin (Saigon), 1960, pp. 4, 67-79, 97-107, and 1965-68, pp. 3, 85-99, 103, 130-39; Far Eastern Economic Review, 1964 Yearbook, pp. 238-40; Food and Agricultural Organization (hereafter cited as FAO) Production Yearbook (Rome: FAO, 1969).

that complementarity of the two economies was rather intensified after the partition, as seen from the pattern of foreign trade presented in Table 1.2. The people of Vietnam, largely engaging in agricultural activities, are grouped around the deltas of two great rivers, the Red River in the North and the Mekong River in the South. The Red River, which originates in Yunnan, China, has a total length of 800 miles, of which about 300 miles cross North Vietnamese territory. The Red River delta covers 5,800 square miles and accounts for over 75 percent of cultivable land in North Vietnam. In terms of surface area, the delta amounts to only about 10 percent, but is nevertheless inhabited by some 77 percent of North Vietnam's population. According to the 1960 census, the delta had a density average of 1,608 persons per square mile, although some places had a density as high as 2,237 (Thai Binh province), 2,091 (Nam Dinh), and 10,408 (Hanoi). In the mountainous region, on the other hand, the density was as low as 16 (Bac Can).[2]

The delta provides the bulk of agricultural produce for the entire Northern population. In 1960, 6.4 million working peasants, or 79 percent of the total labor force, were engaged in agricultural activities, mainly producing rice, the single most important crop.

In the South, the Mekong River delta is the home of 70 percent of the total population. The Mekong delta proper covers only 8,500 square miles, or 13 percent of the total surface area of the South.[3] Population density in the delta ranges from 250 to over 2,000 persons per square mile between the main channel of the Mekong River and Saigon, but its southernmost portions are sparsely settled with as few as 13 persons per square mile. The Mekong delta occupies most of South Vietnam's cultivated land and in peace-time was capable of supplying nearly the entire rice requirement for the whole Vietnamese population.

In 1942 the South was inhabited by 41 percent of the total population of Vietnam, but possessed 59 percent of cultivable land. In a reverse proportion, the North, with 41 percent of arable land, supported 59 percent of the population. In terms of working farmers, the per capita cultivable land was 0.32 hectares for the North as compared with 0.45 hectares per capita for the South in 1960.[4] A 1965 comparison of density of agricultural population of North and South Vietnam with other countries is presented in Figure 1.1.

In terms of paddy output, the 1942 figures show that, of the 6,046,000 tons of total output, more than 61 percent, or 3,711,000 tons, was produced by the South. Of this amount, Cochinchina alone produced 3,179,000 tons, or 53 percent of the total. Compared with the 1942 picture, North Vietnam's economic position vis-a-vis South Vietnam appears to have improved somewhat in 1955, the first year the Democratic Republic of Vietnam was in existence. At that time

TABLE 1. 2

Economic Complementarity: Composition of Foreign Trade in North and South Vietnam in Selected Postpartition Years (value in millions of U. S. dollars)

	1957		1960		1963	
Foreign Trade	Value	Percent	Value	Percent	Value	Percent
Exports						
North Vietnam						
Minerals and manufactures	14. 0	34. 0	35. 5	44. 9	81. 0	86. 6
Handicrafts	2. 3	5. 6	20. 8	26. 4		
Agricultural and forestal products	24. 8	60. 4	22. 7	28. 7	12. 5	13. 4
Total	41. 1	100	79. 0	100	93. 5	100
Total as a percentage of Vietnam total		33. 9		48. 3		55. 1
South Vietnam						
Agricultural and forestal products	68. 8	85. 4	75. 3	89. 1	69. 2	90. 7
Rice	20. 0	24. 8	27. 3	32. 3	35. 7	46. 8
Rubber	48. 8	60. 6	48. 0	56. 8	33. 5	43. 9
Other	11. 7	14. 6	9. 2	10. 9	7. 1	9. 3
Total	80. 5	100	84. 5	100	76. 3	100
Total as a percentage of Vietnam total		66. 1		51. 7		44. 9
Total, Vietnam	121. 6	100	163. 5	100	169. 8	100
Imports						
North Vietnam						
Food and food products	32. 4	32. 6	16. 4	12. 7	26. 5	16. 8
Materials and equipment	66. 3	66. 6	112. 8	87. 3	131. 5	83. 2
Other	0. 8	0. 8	—	—	—	—
Total	99. 5	100	129. 2	100	158. 0	100
Total as a percentage of Vietnam total						
South Vietnam						
Minerals and semimanufactures	60. 8	21. 0	51. 9	21. 6	81. 4	28. 4
Coal	1. 4	0. 5	1. 0	0. 4	1. 0	0. 3
Cement	5. 6	1. 9	7. 0	2. 9	9. 6	3. 4
Iron, steelmill products and metals	32. 2	11. 2	21. 5	8. 9	41. 0	14. 3
Petroleum and petroleum products	15. 3	5. 3	17. 2	7. 2	19. 4	6. 8
Chemicals	6. 3	2. 1	5. 2	2. 2	10. 4	3. 6
Manufactures	107. 3	37. 2	94. 1	39. 1	85. 1	29. 7
Textiles, yarn and thread	56. 3	19. 5	44. 2	18. 4	34. 5	12. 1
Food and tobacco	26. 5	9. 2	26. 5	11. 1	32. 0	11. 2
Other	94. 1	32. 6	67. 8	28. 2	87. 7	30. 7
Total	288. 7	100	240. 3	100	286. 2	100
Total as a percentage of Vietnam total						
Total, Vietnam	388. 2	100	369. 5	100	444. 2	100
Trade balance: surplus or deficit (-)						
North	-58. 4	22. 0	-50. 2	24. 4	-65. 5	24. 0
South	-208. 2	78. 0	-155. 8	75. 6	-209. 9	76. 0
Total, Vietnam	-266. 6	100	-206. 0	100	-274. 4	100

Note: For North Vietnam, data are converted from rubles with the following exchange rates: from 1955-60, 1 ruble = U. S. $0. 25; from 1961, 1 new ruble = $1. 11. For South Vietnam, the pre-devaluation rate: U. S. $1 = VN $35.

Sources: For North Vietnam: estimates based on various DRVN sources. Absolute values (for total exports and imports) are available for 1958-1960. Compositions of trade in percentages are official figures. For 1963, only percentages are given; estimates are based on official data on "growth rates in foreign trade" with 1960 as base year. See Nam Nam Xay Dung Kinh Te va Van Hoa (Hanoi: Central Statistical Office, 1960), especially pp. 203; Vo Nhan Tri, Croissance Economique de la Republique Democratique de la Viet-Nam (Hanoi: Editions En Langues Etrangeres, 1967), especially pp. 318-9, 355-7 and 470; Luu Van Dat, "Ba Nam Hoat Dong Ngoai Thuong," Nghien Cuu Kinh Te (Hanoi), no. 3, (1961); Ly Ban, "Nhiem Vu Cua Nganh Ngoai Thuong Trong Ke Hoach Nam Nam Lan Thu Nhat," Hoc Tap, no. 6, 1963; Ly Ban, "Muoi Nam Phat Trien Ngoai Thuong," Hoc Tap, no. 1 (1965). For South Vietnam: Republic of Vietnam, Institute of Statistics, Nien Giam Thong Ke, 1965-66, 1966-67; AID, Annual Statistical Bulletin (Saigon), no. 9 (1966); National Bank of Vietnam, various data.

FIGURE 1.1

Density of Agricultural Population,
to Cultivable Land, in Selected Countries, 1965

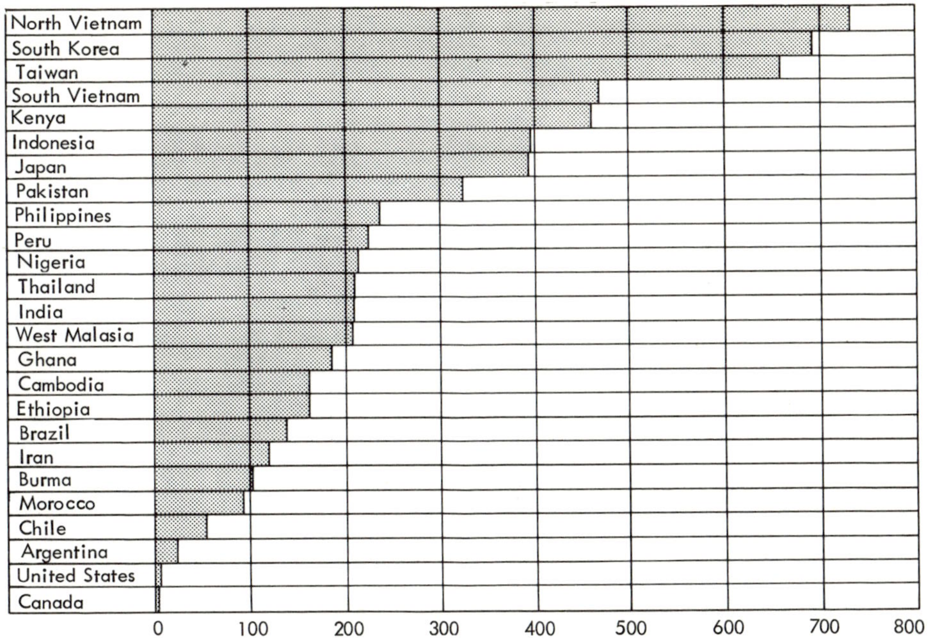

Sources: See page 18.

North Vietnam's share of total population had shrunk from 59 to 53 percent, while cultivated area increased from 41 to 42 percent, and paddy production rose from 39 to 43 percent. This improvement was due to two factors: a large decrease in area under cultivation in Cochinchina during years of warfare; and the departure from North Vietnam to the South of nearly 1 million people, which reduced North Vietnam's 1955 population by 8 percent.[5]

Details of the agricultural sectors of the two Vietnams are provided in Tables 1. 3 and 1. 4, which show area under cultivation, yields, and production on a province-by-province basis in a prepartition year. The figures in Table 1. 2 are significant because, in the absence of accurate statistics since 1955 for North Vietnam, they provide a point of reference that permits a reasonable appraisal of the North's agricultural performance, as land and yield do not change drastically over a short period of time.

Total paddy production in North Vietnam amounted to 2, 335, 000 metric tons in 1942. This had to support a population of 12, 550, 000 persons, implying a per-capita yield of only 186 kilograms. In the same year South Vietnam's production was 3, 711, 000 metric tons for a population of 8, 900, 000 persons, a per-capita yield of 420 kilograms, or more than double the Northern ratio.

Based on the assumption that the average paddy requirement for a Vietnamese is about 240 kilograms per year,[6] the Northern per-capita deficit in 1942 was 54 kilograms as compared with a Southern surplus of 180 kilograms. The discrepancy between the central region (Annam), and the southern region (Cochinchina) was far greater, the latter's per capita output being more than triple that of the former's. On a local basis, per-capita output in North Vietnam varied widely from one province to another, ranging between 340 kilograms (in Bac Giang) and 50 kilograms (in Cao Bang and Quang Yen). Provinces that achieved relatively higher per-capita output were generally those with relatively smaller populations, that is, with higher land/man ratios. It is significant to note that only five provinces (Bac Giang, Ha Dong, Ninh Binh, Phuc Yen, and Vinh Yen) in the entire area above the 17th parallel were self-sufficient in 1942; that is, they had a per-capita output equal to or greater than the minimum yearly requirement.

North Vietnam's rice production is concentrated heavily in the delta provinces, of which three alone (Thai Binh, Hai Duong, and Nam Dinh) accounted for some 38 percent of total output and 30 percent of cultivated land. Relatively high yields were found in the highland provinces such as Son La (3. 7 tons per hectare), Bac Can (2. 0), and Yen Bay (1. 7). Low yields in the delta provinces can be explained in terms of the effects of diminishing returns resulting from the intensive application of labor in that area.

TABLE 1.3

Paddy Output in North Vietnam: Area under Cultivation, Production, Yields, and Per-Capita Output, 1942

Region and Province	Cultivation (thousand hectares)	Production (thousand metric tons)	Yield (tons per hectare)	Population (thousands)	Per-Capita Output (quintals)
Tonkin					
Bac Giang	103	97	0.9	284	3.4
Bac Ninh	83	105	1.3	604	1.7
Ha Dong	109	155	1.4	994	1.6
Hai Duong	82	245	1.3	825	3.0
Ha Nam	61	89	1.5	586	1.5
Hung Yen	85	114	1.3	525	2.2
Kien An	65	87	1.3	429	2.0
Nam dinh	160	214	1.3	1,235	1.7
Ninh Binh	80	125	1.3	424	3.0
Phuc Yen	44	46	1.0	180	2.5
Son Tay	44	70	1.6	321	2.2
Thai Binh	183	250	1.4	1,125	2.2
Vinh Yen	63	80	1.3	290	2.8
Bac Can	6	12	2.0	63	1.9
Cao Bang	12	11	1.0	215	0.5
Ha Giang	10	12	1.2	110	1.0
Hai Ninh	9	8	0.9	109	0.7
Hoa Binh	10	14	1.4	67	2.0
Lai Chau	7	8	1.1	66	1.1
Lang Son	57	22	0.4	204	1.0
Lao Kay	5	9	1.7	59	1.4
Phu Tho	38	34	0.9	335	1.0
Quang Yen	10	9	0.9	185	0.5
Son La	3	11	3.7	106	1.0
Thai Nguyen	34	31	0.9	150	2.0
Tuyen Quang	10	9	0.8	82	0.9
Yen Bay	11	13	1.7	100	1.9
Total or average[a]	1,487	1,882	1.3	9,920[b]	1.9
Rice equivalent		1,223			1.2
Northern Annam					
Thanh Hoa	178	224	1.2	1,107	2.0
Nghe An	121	106	0.9	700	1.5
Ha Tinh	97	98	1.0	573	1.7
Quang Binh	42	25	0.6	250	1.0
Total or average	438	453	0.9	2,630	1.7
Rice equivalent[a]		294			1.1
Total or average in North Vietnam	1,925	2,335	1.2	12,550	1.9
Rice equivalent[a]		1,517			1.2

[a]Based on paddy-to-rice conversion rate of 0.65 (one kilogram of unprocessed rice, or paddy, gives an average of 0.650 kilograms of rice, afterhusking).

[b]Including population in areas not shown here.

Note: Sums may not add to total because of rounding.

Source: Union Francaise, Haut Commissariat de France pour l'Indochine, *Annuaire Statistique de l'Indochine*, vol. X, 1941-42, pp. 87-88.

TABLE 1.4

Paddy Output in South Vietnam: Area under Cultivation, Production, Yields, and Per-Capita Output, 1942

Region and Province	Cultivation (thousand hectares)	Production (thousand metric tons)	Yield (tons per hectare)	Population (thousands)	Per-Capita Output (quintals)
Cochinchina					
Bac Liêu	318	309	1.0	308	10.0
Bà Rịa	13	16	1.3	65	2.5
Bến Tre	105	185	1.8	341	5.4
Biên Hoa	28	26	0.9	194	1.3
Cần Thơ	189	340	1.8	420	8.1
Chau Doc	97	107	1.1	271	4.0
Cho Long	97	144	1.5	278	5.2
Gia Dinh	55	85	1.5	341	2.5
Gò Cong	50	101	2.0	118	8.6
Hà Tiên	6	6	1.1	29	2.1
Long Xuyen	161	259	1.6	272	9.5
My Tho	140	221	1.6	417	5.3
Rach Gia	382	392	1.0	377	10.0
Sa Dec	92	139	1.5	258	5.4
Soc Trang	203	310	1.5	237	13.0
Tan An	73	79	1.1	155	5.1
Tay Ninh	55	52	0.9	143	3.6
Thu Dau Mot	15	17	1.2	216	0.8
Tra Vinh	145	243	1.7	281	8.7
Vinh Long	79	148	1.9	219	6.8
Total or average	2,303	3,179	1.4	5,400[a]	5.9
Rice equivalent		2,066			3.8
Southern Annam					
Binh Dinh	110	104	0.9	727	1.4
Binh Thuan	58	69	1.2	134	5.1
Khanh Hoa	47	43	0.9	144	2.9
Ninh Thuan	11	19	1.7	108	1.7
Phu Yen	53	51	1.0	274	1.8
Quang Nam	86	85	1.0	989	0.9
Quang Ngai	59	54	0.9	478	1.1
Quang Tri	29	36	1.2	190	1.9
Thua Thien	55	71	1.3	399	1.8
Total or average	508	532	1.0	3,500[a]	1.5
Rice equivalent		346			1.0
Total in South Vietnam	2,811	3,711	1.3	8,900	4.2
Rice equivalent		2,412			2.7

[a]Including population in areas not shown here.

Note: Figures may not add to total because of rounding.

Source: Union Francaise, Haut Commissariat de France pour l'Indochine, Annuaire Statistique de l'Indochine, vol. X, 1941-42, pp. 87-88.

With the exception of Thanh Hoa, where land is fertile, the agricultural picture of North Vietnam's southern provinces was similar to that of the other part of Annam, the poorest region of the whole country. In 1942 the average yield for those provinces was 0.9 tons per hectare compared with the average of 1.3 tons for the whole of the North. Per-capita production was as low as 100 kilograms in Quang Binh province and as high as 200 kilograms in Thanh Hoa province.

In the same year Cochinchina presented a strikingly different picture from the rest of the country. Although the productivity of land is not much higher than that of Tonkin, a larger area under cultivation and a smaller population give a per-capita output of an average of 590 kilograms, ranging from between 1.3 tons (in Soc Trang) and 1.0 ton (Bac Lieu and Rach Gia) to 80 kilograms (Thu Dau Mot). The above comparison between North and South is based on agricultural performance in all Vietnam in a relatively good year.

The economic history of North Vietnam shows, however, that bad years are more the rule than the exception and that agricultural instability is among the most serious obstacles to its economic and social development. In order to have a better grasp of North Vietnam's economy, therefore, an elaboration of its agricultural fluctuations is warranted.

NORTH VIETNAM'S ECONOMIC INSTABILITY

The major sources of economic instability in North Vietnam are adverse climatic conditions and the violence of the Red River's flow. Northern weather is well known for its extremes—too much rain or too little causing either floods or drought.

During the rainy season, from May to September (or late May to October), the climate is characterized by a very high degree of precipitation. Rainfall is propitious for agriculture, but extreme violence and irregularity constitute a threat of floods, water logging, and erosion. The summer rains are usually storm rains. In a few hours a storm may bring massive quantities of water.[7] Heavy rains are often accompanied by great winds that turn into typhoons with a velocity reaching speeds over 100 miles per hour.[8] In September 1927 a typhoon at Phu Lien brought 41 inches of water in four days, destroying nearly the entire crop.[9] Typhoons come from the China Sea and take place mostly between July and November, there were 147 typhoons between 1911 and 1922,[10] that is, on the average, nearly eight per year.

Drought is another problem in North Vietnam, although not as serious as floods. Even during the rainy season, there are short dry periods that are very harmful to the rice fields at high grounds where irrigation is difficult. During the winter months (November-April), drought accompanied by a high degree of humidity may endanger the summer crop. Nevertheless, as Pierre Gourou pointed out, these droughts are not generally catastrophic since they tend to diminish the yields of the crop rather than destroy them as floods do.[11]

In addition to the climate factor, North Vietnam's agriculture is very vulnerable to the violence of the Red River's flow. Wet-rice cultivation in Vietnam depends heavily on water, the leading input. A successful crop requires not only an adequate supply of water, but also its timely application, its level at different stages of growth of the rice plant. While the Red River supplies water for paddy sowing, transplanting, and growth, it also floods to destroy the final output.

The Red River is often compared with the Mekong River in the South; while the latter is calm and flows through the South as an irrigator (like the Nile) of adjoining lands, the former, in Gourou's words, "crosses the delta as a fearful stranger rather than as a helpful collaborator."[12] It is so strange and fearful that for centuries it has never ceased to be a source of greatest anxiety to the peasants and their rulers, be they kings, foreign powers, or the Communist regime. During the summer its flow may reach 800,000 cubic feet per second. The period of floods occurs from around June to October.[13] The great danger of the floods stems from the fact that they take place in the same season when the water level is normally high. The enormous and sudden flow of water is accentuated by the surrounding mountains; masses of water gather momentum and increase in volume during the journey downstream. Then they threaten man and his entire crop and "may cover about the whole delta if they were not contained,"[14] since most of the delta is no more than ten feet above sea level.

The economic life and hence the social and political stability of the delta depend to a large extent on an elaborate system of dikes to control the river and to hold back the sea. The present system of nearly 3,000 miles of primary, secondary, and tertiary dikes is the product of thousands of years of work. History does not record exactly when the first dikes were built, but fragmentary information suggests that they had already existed at the dawn of the Christian era. When the Chinese occupied the land at about the second century B.C., they began to strengthen and extend existing dikes. Regular inspection of the dikes by members of the Mandarin courts was frequently carried out.[15]

Under the French administration (1884-1945), the embankments were raised above the high-water level of the Red River, and several modern water-conservation works were built. Administrators at all levels in each province, from the French resident, the <u>tong doc</u>

(province chief), down to the ly truong (village chief) were responsible for defending the dikes.

In spite of all efforts, the prospect of completely defending the dikes has never appeared to be very good. From 1890 to 1926 Tonkin suffered 16 major breakings of dikes (see the appendix to this chapter).

The agricultural sector in North Vietnam is such that even under normal conditions it is not capable of supporting the Northern population. Traditionally, therefore, the North has had to count on the South to make up for its annual rice deficit. In years of great calamities, that deficit became critical and further aggravated the overall food problem. Because of this dependence on the South for food, North Vietnam has encountered great difficulties during the times when the Southern foods were not available.

The tragedy that took place in 1945 is more than sufficient to illustrate the gravity of this dependence. In that year, mass starvation took place in North Vietnam, resulting in the loss of from 1.5 to 2 million lives, or about 17 percent of North Vietnam's population at that time.[16] The crisis was so serious that it affected the very center of Hanoi.[17] It was brought about when major floods took place at the time when normal economic relations between North and South were discontinued because of World War II.

A very prolonged flood in August-September 1944 had already destroyed a major part of seedlings and reduced the tenth-month crop of that year by 30 percent. This was followed by a more serious flood in 1945, which resulted in the breaking of a great number of Red River dikes. Describing the flood situation in that year, a noted observer of Vietnamese affairs wrote, "When Ho Chi Minh travelled along the road from Thai Nguyen on the way to Hanoi in August 1945, he passed through ravaged lands. Roads were under water and ricefields had become lakes which seemed to have no shores."[18]

It was not possible at that time for the North to obtain relief from the South in order to avert the crisis, because of the breakdown of the transport network. Furthermore, the rice available in the South was also reduced because of the Japanese and Vichy French policies regarding collection and storage of foodstuffs for reserves.[19]

CONCLUSION

Indeed, the 1945 tragedy serves well to illustrate the socio-politico-economic realities of the two Vietnams. First, it shows the vulnerability of North Vietnam's economy; second, it dramatizes the degree of North Vietnam's dependence on the South for food; and third, it illustrates the far-reaching political consequences that can accompany a major food crisis. One of the principal consequences of the famine in that year was the success of the Vietminh revolution. The starvation and suffering no doubt rendered the climate most favorable for igniting the August 17-18 revolution and thus fostering the birth of the Democratic Republic of Vietnam (DRVN) on September 2. Amidst widespread starvation the Communist party launched an intensive propaganda effort centering around the food issue and based on an unqualified promise to solve the problem of the rice deficit and to bring about radical land reforms. The promise was backed up by distribution of large quantities of rice seized from the Japanese storage houses toward the end of 1945.

Tables 1. 1, and 1. 2 which shows the distribution of resources and the structure of foreign trade in the two Vietnams, indicate a high degree of complementarity of the two economies. But if the economic relationship arising from this complementarity is beneficial to South Vietnam, it is essential to North Vietnam. Because of the instability of its annual agricultural output, North Vietnam depends heavily on South Vietnam to compensate for its food shortage.

Commenting on the economic relationship of the two Vietnams, a noted French author, Jean Lacouture, put it this way:

> Before the partitioning of 1954 Viet-Nam certainly had not always been politically united, and the division of the colonial epoch—with Tonkin in the North, Annam in the Center, and Cochinchina in the South—corresponded quite well to the realities of cultural traditions and collective psychology. But in the economic domain, Vietnamese unity took on overriding importance. . . . To create two political capitals, Hanoi and Saigon, was partially justified, but to break up Viet-Nam's economic unity was to attempt the irrational.[20]

In order to provide a better understanding of the initial conditions of North Vietnam's economy, therefore, the next chapter analyzes the economic relationship between the two zones prior to partition.

APPENDIX: RECORDS OF DIKE BREAKINGS*

The history of the dike system in North Viet-Nam is a history of dike breaking. Between 1890 and 1926 the Tonkin delta suffered

---

*Excerpted from the author's "The Dikes in North Vietnam," New York Times, International Edition, and International Herald

sixteen major breakings of dikes (1890, 1893, 1894, 1896, 1902, 1903, 1904, 1905, 1909, 1911, 1913, 1915, 1918, 1923, 1924, and 1926): that is, on average, there was one major breaking for about every two and a quarter years. In 1915 the dikes broke at 48 places and flood waters submerged 365,000 hectares, or one-third of the area under cultivation.

Between 1927 and 1945, in spite of new, more modern works constructed under the Frend administration, the dikes broke on seven occasions: 1930, 1936, 1937, 1941, 1942, 1944, and 1945. The 1945 calamity was the worst of all, submerging 250,000 hectares in nine provinces, destroying most of the winter rice crop and resulting in a widespread famine.

Since 1954 North Viet-Nam has not had much luck in controlling seasonal floods. Perhaps with the exception of 1956, 1959, 1964, 1965, and 1967, most of the other years have been classified as bad years, during which typhoons, floods and drought dominated.

Natural calamities were climaxed by the devastating flood of 1971.

According to statements from the North Vietnamese leaders, the 1971 flood was one of the most devastating in history, even bigger than that of 1945. On September 2, 1971 Premier Pham van Dong publicly declared that: "Flood waters from upstream have caused the water in the entire system of the Red River and Thai Binh River to rise to an unprecedented level, while heavy downpours in the delta caused added difficulties for the defense of the dike system."

Under the Communist regime, the dikes occupy the center of its agricultural policies. From 1955 to 1964, work on the dikes and irrigation system required 777 million work days, 915 million cubic meters of earthwork, 1.7 million cubic meters of stone and 335,000 cubic meters of concrete. The average state expenditures on hydraulic works accounted for over 70 percent on total annual spending on agriculture. Each peasant is called upon to devote between 60 and 120 days each year to labor on public works for the protection of the fields.

In spite of all efforts, the prospect of completely defending the dikes appears to be as remote now as ever before.

Tribune, September 7, 1972. For data on dike breakings, see Tran Dang Khoa, "Les Problemes Hydrauliques au Nord Vietnam," in Democratic Republic of Vietnam (DRVN), Etudes Vietnamiennes, no. 2 (Hanoi: Editions En Langues Etrangeres, 1964), p. 72.

NOTES

1. Paul Bernard, La Probleme Economique Indochinois (Paris: Nouvelles Editions Latines, 1934), pp. 185-86.

2. Democratic Republic of Vietnam (hereafter cited as DRVN), Central Statistical Office, Nam Nam Xay Dung Kinh Te Va Van Hoa (Five years of economic and culture building) (Hanoi: Central Statistical Office, 1960), pp. 65-66.

3. For details on the Mekong River and its basin, see C. Hart Schaaff and Russell H. Fifield, The Lower Mekong: Challenge to Cooperation in Southeast Asia (Princeton, N. J. : Van Nostrand, 1963), pp. 71-79.

4. The comparison of per-capita cultivable land for working farmers in North and South Vietnam is based on 1960 data because of their availability. North Vietnam's cultivable land was 2,036,000 hectares, and its population of working farmers was 6,377,024 persons (DRVN, op. cit., p. 70). South Vietnam's cultivable land was 2,471,000 hectares, and its population of working farmers was 5,450,000 (U. S., Agency for International Development Annual Statistical Bulletin [Saigon], no. 9 [1966], p. 10). Per-capita arable land elsewhere in Southeast Asia was 0.9 hectares in Thailand, 1.2 hectares in India, and 1.2 hectares in the Philippines. See, United States Department of Agriculture, "Changes in Agriculture in 26 Developing Nations, 1948-1963," Foreign Agricultural Economic Report no. 27, 1965.

5. Reduction of Cochinchina's cultivated land: In 1954, the cultivated area in Cochinchina was only 1,572,000 hectares and in 1955, only 1,803,000 hectares, amounting respectively to 68 percent and 78 percent of the area under cultivation in 1942. See Republic of Vietnam (hereafter cited as RVN), Ministry of Agriculture, Agricultural Statistics Yearbook, 1965 (Saigon, 1966), p. 34. Reduction of North Vietnam's 1955 population: According to North Vietnam's official data, the population in 1955 was estimated at 13,574,000 persons, of which the rural population was 12,577,000, or 92.6 percent, and the urban population was 997,000, or 7.4 percent (DRVN, op. cit., p. 71).

6. The 240-kilogram per-capita figure is considered as the standard requirement by the National Institute of Statistics, by the Vietcong as a base for tax deduction, and by others such as Colin Clark and Margaret Haswell. See Robert L. Sansom, The Economics of Insurgency in the Mekong Delta of Viet-Nam, (Cambridge: MIT Press, Massachusetts, 1970), pp. 37-38 and footnotes thereto.

7. For details on this point, see Pierre Gourou, The Peasants of the Tonkin Delta, A Study of Human Geography, vol. 1 (New Haven: Human Relations Area Files, Inc., 1955), pp. 61-64.

8. Speeds were recorded at Phu Lien and Lan Son on August 25, 1924. For more details, see Tran Dang Khoa, "Les Problemes Hydrauliques au Nord Viet-Nam," in DRVN, Etudes Vietnamiennes, no. 2 (Hanoi: Editions En Langues Etrangeres, 1964), pp. 46-68.

9. Gourou, op. cit., p. 66.

10. Khoa, op. cit., p. 47.

11. Gourou, op. cit., p. 71.

12. Ibid., p. 76.

13. For an engineering study of the Red River, see M. Parde, "Le Regime du Fleuve Rouge," in Annales de Geographie (Paris: Befeo, 1938), especially pp. 191-95.

14. Gourou, op. cit., p. 77.

15. For a description of dike supervision and penalty on failure to comply under various dynasties, see M. R. Deloustal, La Justice dans l'Annam, (Paris: Befeo, 1910), p. 795.

16. Bui Huy Dap, "Nhung Bien Doi Cach Mang Sau-Saco Nong Thon va Nong Nghiep Mien Bac" (Profound revolutionary changes in rural area and in agriculture in North Vietnam) in Khoa Hoc va Ky Thuat Nong Nghiep (Science and Agricultural Technology), (Hanoi), no. 2 (1970): 70; Le Chau, Le Viet-Nam Socialiste, Une Economie de Transition (Paris: Francois Maspero, 1966), p. 58 (hereafter cited as Le Viet-Nam Socialiste). Other estimates are from 400,000 to 2,000,000 (Alexander Woodside, "Decolonization and Agricultural Reform in Northern Viet-Nam," Asian Survey, [University of California, Berkeley], 10, no. 8 [August 1970]: 706).

17. For a detailed description of the crisis, see Francoise Martin, Heures Tragiques au Tonkin (Paris: Berget-Levroult, 1948), pp. 99-105.

18. Ellen J. Hammer, The Struggle for Indochina 1940-55 (Stanford: Stanford University Press, 1967), p. 145.

19. See Girard Chaliand, The Peasants of North Viet-Nam (Middlesex, England: Penguin, 1969), pp. 34-5.

20. Jean Lacouture, Vietnam: Between the Two Truces (New York: Vintage, 1966), pp. 36-37.

FIGURE SOURCE NOTE

FAO, Production Yearbook (Rome: FAO, 1966) Tables 1, 4, 5; United Nations, Statistical Yearbook (New York, 1966) Table 17, reprinted in Land Reform in Vietnam, Vol. 4, pt. 1 (Menlo Park, Calif.: Stanford Research Institute, 1968), p. 12.

CHAPTER

2
STRUCTURE OF INTERZONAL TRADE

No study has so far been made to determine the structure of intra-Vietnam economic relations prior to partition, and available data bearing on the question are extremely fragmentary. Nevertheless, based on the economic complementarity of the two regions and on the free movement of commerce and labor before World War II, the level of trade between the two Vietnams in relation to their production and consumption appears to have been high in comparison with the level of intraregional trade in other parts of the world.[1]

It must be kept in mind, however, that it would be erroneous to consider this flow the same way as one would consider a volume of trade between two foreign countries, since the latter involves tariff barriers as well as currency and payment problems. The flow of commerce within one country is spontaneous, often determined by the degree of regional specialization and the transport facilities.

While it is not possible to present an accurate analysis of the complete structure and pattern of economic interchange between the different regions of Vietnam, some important components of this exchange are illustrated in Tables 2.1 and 2.2. The tables show only some magnitude of the trade flow. Because they are derived from different sources, they are subject to error. Data in the two tables are also reclassified from original data, related to the three parts of Vietnam, to fit the present-day division between North and South. Furthermore, they represent only the volume of controlled trade, that is, the part of merchandise transported mainly by sea and railway between the points of controlled loading and unloading. As such, the figures exclude goods transported by road and those that went through uncontrolled trade channels. In regard to paddy, shipments in the years between 1951 and 1954 represent only the required supply to accommodate the needs of the portion of the population under French control in North Vietnam, which was only about one-half of the total Northern population at the time.[2] It may be concluded, therefore, that the level of real trade flow was significantly higher than that shown in the tables.

TRADE FLOW FROM NORTH TO SOUTH

The North is well known for the richness of its subsoil. The distribution of production in major commodities (in Table 1. 1) shows concentration of mining resources in the North, which produced 100 percent of Vietnam's coal output in 1942. The coal is largely anthracite of high quality (relatively smokeless with low ash content), and is North Vietnam's most important mineral; together with zinc and tin production, it amounted to 96. 5 percent of total mineral production in the 1930s. Other less important minerals are iron ore, lead, phosphate, and gold. Some 90 percent of Indochinese mines are located in Tonkin (the rest are in Laos). The mines, situated near seaports and railways, provide employment for the bulk of Northern skilled workers in peacetime.

Coal

Coal represents the greatest share in the North-South trade flow. The Quang Yen coal basin deposits, composed mainly of the Hon Gay and Dong Trieu mines, are among the largest coal beds in Southeast Asia. Prior to the French conquest in 1884, the coal mines had long been exploited by the Chinese, though on a very small scale. Systematic exploration was begun in 1881-82 by E. Fuchs of the School of Mines in Paris. This was followed by extraction activities by two French companies, the Societe des Charbonnages du Tonkin and the Societe des Charbonnages du Dong Trieu.[3] Speculation had it that the richness of the Northern mines was among the main factors that attracted French interest in conquering Indochina.[4]

Coal production increased rapidly in the early 1900s, reaching a record level of 2. 6 million tons in 1939 as compared with 501, 000 tons in 1913. Although Northern coal was extracted mainly for export outside Indochina, an important part of the coal production was shipped to the South. In 1939, for example, about 190, 000 tons, or 10 percent of exported coal, went to the South. Total export in that year was 247, 000 tons, accounting for 81 percent of total output (see Table 2. 1). Production declined sharply during the 1940s because of Allied bombing during World War II and deteriorating security conditions thereafter. By 1950, coal output dwindled to 503, 000 tons or just slightly higher than the 1939 level. Production

TABLE 2.1

Trade Flow from North to South in Relation to
Production and Total Exports in Selected Prepartition Years

Product and Year	Production	Exports	To South Vietnam: Quantity	To South Vietnam: Percentage of Total Exports	To Rest of World: Quantity	To Rest of World: Percentage of Total Exports	Exports as Percentage of Production
Coal (thousands of tons)							
1939	2,615	1,981	190	10	1,791	90	68
1950	503	181	122	67	59	33	36
1951	639	342	126	37	216	63	53
1952	895	357	144	40	213	60	40
1953	887	387	81	21	306	79	44
1954	944	571	110	19	461	81	60
Cement (thousands of tons)							
1939	306	247	90	36	157	64	81
1950	144	95	78	82	17	18	66
1951	212	109	97	89	12	11	51
1952	225	100	95	95	5	5	44
1953	291	143	142	99	1	1	49
1954	225	130	130	100	0	0	51

Note: North and South Vietnam are as defined by the 1954 Geneva agreement.

Sources: Republic of Vietnam, Service de la Statistique Generale, Annuaire Statistique du Vietnam, vol. 1, 1949-50, pp. 48-50, 116, 117, 128, 134, 135, 139, 257, vol. 2, 1950-51, pp. 116, 132, Institute d'Emission des Etats du Cambodge, du Laos, et du Vietnam, Compte Rendu des Operations (Saigon, 1954), pp. 91, 96; Union Francaise, Haut Commissariat de France pour I'Indochine, Annuaire Statistique, vol. 10, 1941-42, pp. 33, 188, 272, 278-9, 283, vol. 11, 1943-46, p. 277, vol. 12, 1947-48, p. 113; United Nations, Toward the Economic Development of the Republic of Vietnam (New York: United Nations, 1959), p. 33.

gradually increased in the early 1950s, reaching 944,000 tons by 1954, almost double the 1950 figure. As Indochina's trade with foreign countries shrank during 1950-54, the North-South coal trade boomed. In 1950, for example, the share of coal shipped to the South in relation to total production reached 67 percent; it was 37 percent in 1951 and 42 percent in 1952.

The role of Northern coal in the economy of Cochinchina becomes apparent when viewed in terms of the latter's total coal imports. In 1939 Cochinchina imported from foreign countries (mainly Japan and India) a little over 31,000 tons of coal, or only about 18 percent of the 190,000 tons received from Tonkin. In 1950 and 1951 Northern shipments were 122,000 tons and 126,000 tons, respectively,[5] or more than ten times higher compared with total imports from other countries of 11,380 and 11,964 tons. From 1952 to 1954, the situation remained roughly the same as existed in 1950-51.[6]

Since the partition of the country, the South has been cut off from its traditional fuel supply and has had to increase its imports from abroad.[7]

Cement and Other Products

The availability of coal as a source of fuel permitted North Vietnam to undertake other important industrial enterprises; among them, the cement industry was the most important. The cement plant at Hai Phong, one of the biggest factories in all Indochina, was founded in 1899. It benefited from an abundant supply of raw materials (coal, limestone, and clays) nearby and was located conveniently near the port. The quality of North Vietnam's cement was highly regarded all over the Far East.[8]

Cement production was destined mainly for the Indochinese domestic market. Thus, cement was shipped to the South in high proportions in relation to total export. Of 247 tons exported in 1939 (81 percent of total output), about 90,000 tons, or 36 percent, went to the South. During 1950-54, nearly the entire amount of Northern exported cement was direct to the South. Total exports amounted to 78,000 tons (66 percent of output) in 1950, and 130,000 tons (51 percent of output) in 1954. The volume of South Vietnam's cement imported from foreign countries in those two years was 7,000 tons and 55,000 tons, respectively.

Other mining products reaching the Southern markets from the North included tin, zinc, iron ore, phosphate, apatite, silver, and gold. In 1942 Tonkin produced 99,000 tons of apatite, 16,000 tons of zinc, 500 tons of tin, 13,000 tons of bauxite and—together with Annam—62,000 tons of phosphate, 63,000 tons of pig iron, and 1,438 tons of manganese.[9] It is not possible to measure the size of trade in these materials. However, since only about half of total production of these materials was consumed in North Vietnam, it may be concluded that an important part of the other half was exported to South Vietnam, which did not produce any of these minerals at that time.

In addition to mineral products, the South also imported from the North textiles, glassware, matches, and a variety of handicrafts such as brass and copper products and pottery. In 1951 textile production of the two major Northern companies amounted to 3,210 tons of yarn,[10] and one Northern firm produced 5,932 tons of glass, 12,146,000 bottles, and 36,330,000 matchboxes.[11] The glass and glass products were used in the large-scale Saigon beer and beverage industry.

Movement of Workers to the South

The traditional movement of Northern workers into the South has a special role in the economic relationship between the two regions.

It would be useful in explaining this movement to point out at the outset that the history of Vietnam has one similarity to that of the United States. Both involved continuous migration—westward in the United States, southward in Vietnam. Whereas gold and land were the moving forces in the United States, in Vietnam, it was food and land. To a North Vietnamese, the image of the South is always that of hope, of life. The river there is calmer and the land more bountiful.* Because of the poverty of land and population pressure in the North, the Northern peasants, in spite of their extremely strong attachment to their villages and ancestors' tombs, continued to move southward.

The Northern migration movement began nearly 800 years prior to the French occupation, that is, from the eleventh to eighteenth centuries.[12] During the colonial era, systematic movement of labor from the North was organized under a contracts system

---

*The Northern peasant even considers the wind that comes from the South to be pleasant in contrast to the wind that comes from the North, which is violent and cold. Thus it is considered desirable to expose one's house to the South, as an old saying urges: "To build your house facing the South is as natural as for a man to marry a woman who is gentle and wise" (Lay vo dan ba; Lam nha huong nam). The peasant also bases his agricultural activities on various meteorological principles derived from experience. One important saying gives him a warning signal when he sees the clouds

put into use in 1898 and 1910. Under this system, a laborer, or coolie as he was called by the French, was recruited by French agents to work on rubber plantations in the South with the contract covering a period of from one to three years.[13] The agent received from ten to twelve piasters for each coolie contracted, normally an easy job to undertake. At times when the Tonkin delta was hit by serious floods, the agent might even collect fees from the workers for contracting them to the South.

The economic effect of this continuing migration was on the whole beneficial. In leaving the overcrowded Tonkin delta, the migrants not only provided some relief for its demographic problem, but they also sent back to their families regular income earned on the Cochinchinese plantations. In the South the Northern workers made up for an important part of the shortage of skilled laborers there.

The annual outflow of workers from the North to the South and to the rest of the French Union in relation to total labor force was very high.[14] The size of the labor force may be measured in terms of the number of workers engaged in mining activities since mining dominates North Vietnam's industrial sector, and data are most readily available for this industry.[15]

Table 2. 2 shows the magnitude of migration during 1941-43, a period when economic interchange between the North and the South was still unrestricted. In 1943 the number of workers leaving Tonkin amounted to 21,738 as compared with 13,980 in 1941 and 17,610 in 1942. This represented an average increase of nearly 24 percent per annum. In relation to the remaining labor force (in the mining sector), total outflow was equivalent to 62 percent in 1943 against 28 percent in 1941.[16]

Of the total outflow in 1941, 32 percent were sent to the Southern plantations, the rest to French Pacific islands (New Caledonia, New Hebrides, and so on); only a very small number went to Laos and Cambodia. In 1943 outflow reached 6,920 persons, an increase of 66 percent over the 1941 level.

Figures on the outflow are calculated on a net basis; that is, they were arrived at by subtracting the number of returning North Vietnamese workers from the number of departing ones. Details of the two-way movement are shown in Table 2. 2. The annual outflow on the basis of the table was much higher than the net outflow.

---

approaching from the East while telling him to relax when they come from the South: Con dang dong vua trong vua chay, con dang nam vua lam vua choi ("When clouds come from the East, look while running, but when from the South, play while working").

TABLE 2. 2

Movement of Workers between North Vietnam and Other Parts of the French Union

Country Where Workers Went or Returned From	1941			1942			1943		
	Male	Female	Total	Male	Female	Total	Male	Female	Total
Workers departing									
South Vietnam	7,270	1,170[a]	8,440	9,059	1,681[a]	10,740	8,922	1,447[a]	10,369
Cambodia	4,150	770	4,920	4,902	702	5,604	3,804	783	4,587
Laos	1,580	—	1,580	370	—	370	—	—	—
Other[b]	—	—	15,900	—	—	16,550	—	—	18,200
Total			30,940			33,264			33,156
Workers returning									
South Vietnam	3,730	650	4,380	3,548	618	4,166	3,155	294	3,449
Cambodia	3,220	860	4,080	3,067	101	3,168	1,798	71	1,869
Laos	—	—	—	120	—	120	—	—	—
Other[b]	—	—	8,500	—	—	8,200	—	—	6,100
Total			16,960			15,654			11,418
Net outflow of workers from North			13,980			17,610			21,738
Number of workers in Northern mines			49,600			44,300			35,000
Net outflow as percentage of number of workers			28			40			62

[a]Including about 400 children.

[b]Including other territories of the franc area and French Pacific islands.

Note: Data include contracted and full-time workers only; the workers are mainly from Tonkin and from the three northern provinces of Annam, which are now a part of North Vietnam's territory.

Sources: Republic of Vietnam, Institute of Statistics, Annuaire Statistique de l'Indochine, vol. 10, 1941-42, pp. 28, 278, 284, and vol. 11, 1943-44, pp. 33, 272, 279.

It amounted to 33,156 persons in 1943 as compared with 30,940 in 1941, averaging some 82 percent higher than the figures for net outflow. Thus, in 1943 the size of the work force that actually left North Vietnam equaled roughly the size of the work force in the mining sector. During 1941-43 the number of workers who went to South Vietnam increased by 21 percent, while the number of those who returned from the South declined by 27 percent suggesting that more and more workers who left North Vietnam did finally settle in the South.

During the same period, the rate of return of workers from Cambodia and the Pacific islands also declined, even at a much faster rate (54 and 72 percent, respectively) indicating that once most Northern workers arrived in Cambodia and other parts of the French Union, they tended to remain there.[17] As the journey from the North to the South was considered a hazardous adventure, most of the departing workers were male rather than female.[18]

The size of the outflow of Northern workers discussed above, although high, represented only those who were recruited under the contract system. In addition to the recorded flow of workers, there was an unknown seasonal movement of laborers who went South to seek temporary work on the plantations as well as of those skilled workers seeking employment in urban areas. They were carpenters, bricklayers, blacksmiths, artisans, lacquerers, wood-carvers, tailors.

During the colonial era, the French wanted to encourage a higher and more permanent migration from the Tonkin delta. The argument was that "even at a primitive level of technology in agricultural farming, the rice fields need only two persons to cultivate one hectare, and on this basis, North Vietnam had three times the number of farmers which it needed for cultivation."[19]

TRADE FLOW FROM SOUTH TO NORTH

Foodstuffs

The most important item entering interzonal trade prior to partition was foodstuffs, largely rice. As was shown in Chapter 1, approximately 61 percent of Vietnam's rice land and production are in the Southern part of the country. In the 1930s Cochinchina was among the leading rice exporters in the world market, ranking behind only Burma and Thailand, Cochinchina exported around 1.5 million tons of rice annually. In addition to exporting to foreign countries, it supplied rice to the rest of Indochina, especially to the Tonkin delta. Fragmentary information from both Communist and free-world sources indicates that before World War II, about 200,000 to

TABLE 2.3

Trade Flow from South to North in Relation to
Production and Total Exports in Selected Prepartition Years

Product and Year	Production	Exports	To North Vietnam: Quantity	To North Vietnam: In Percentage of Total Exports	To Rest of World: Quantity	To Rest of World: In Percentage of Total Exports	Exports as Percentage of Production
Rice (thousands of tons)[a]							
1939	2,900[b]	1,903	250[b]	12	1,673	88	66
1942	2,543	1,224	250[b]	20	974	80	48
1950	1,036	221	110	50	111	50	21
1951	1,249	356	81	23	275	77	29
1952	999	326	96	29	230	71	33
1953	1,003	368	171	46	197	54	37
1954	1,043	520	167	32	353	68	50
Rubber (tons)[c]							
1939	67,500	—	—	—	47,541	—	—
1946	12,921	94,682	167	0.2	94,515	—	—
1947	26,196	35,813	135	0.4	35,678	—	—
1948	27,633	29,606	581	2.0	29,025	—	—
1949	27,386	28,797	139	0.5	28,658	99.5	105[d]
1950	32,968	36,267	128	0.4	36,139	99.6	110
1951	37,280	37,109	142	0.4	36,967	99.6	99

[a]Figures for Cochinchina only. Production and export figures are in tons of rice. The conversion rate from paddy to rice is 0.65. Export figures for 1939-42 include some from Cambodia (about 8 percent).

[b]Estimated. For 1939 rice production: 2.3 million hectares and estimated yield of 1.26 tons per hectare

[c]Official data on exports were available for all Indochina, thus include rubber shipments from Cambodia and Laos. In prepartition time, however, South Vietnam's production accounted for about 69 percent of total output of Indochina; therefore, quantity was calculated on basis of 69 percent of total Indochinese exports.

[d]Exports were higher than production because they include stocking from previous years.

Sources: Republic of Vietnam, Service de la Statistique Generale, Annuaire Statistique du Vietnam, vol. 1, 1949-50, pp. 48-50, 116, 117, 128, 134, 135, 139, 257, vol. 2, 1950-51, pp. 116, 132; Institute d'Emission des Etats du Cambodge, du Laos, et du Vietnam, Compte Rendu des Operations (Saigon, 1954), pp. 91, 96; Union Francaise, Haut Commissariat de France pour l'Indochine, Annuaire Statistique, vol. 10, 1941-42, (pp. 283, 33, 188, 272, 278-9, vol. 11, 1943-46, p. 277, vol. 12, 1947-48, p. 113; United Nations, Toward the Economic Development of the Republic of Vietnam (New York: United Nations, 1959) p. 33.

250,000 tons of paddy from the Mekong delta were sent to North Vietnam each year.[20]

Although data are not available regarding the total flow of trade from South to North, Table 2.3 shows partial data on rice and rubber exported to North Vietnam during 1950-54 in relation to production and exports of South Vietnam.

From an estimated 250,000 tons, or 12 percent of total exports, rice supplied to North Vietnam fell to 81,000 in 1951 (23 percent of total exports), then increased again to 171,000 and 167,000 tons in 1953 and 1954, respectively. The sharp reduction in quantity was due to the breakdown of the North-South railway network.

During the five years preceding partition, Cochinchinese rice shipments to North Vietnam in relation to total exports were very important, accounting for between 23 percent (1951) and 50 percent (1950), as compared with about 12 percent in 1939. Table 2.4 shows details of the controlled rice flow during 1950-54. Total shipments in that period ranged from 81,419 tons in 1951 to 171,193 tons in 1953, averaging 125,000 tons per annum.

Available data for a quarterly breakdown of rice shipments in 1953 and 1954 indicate that the largest shipments normally took place just before the harvest in the North, when rice reserves were low. In 1953, for example, 32 percent of the year's shipment arrived at the Northern market during the rice-transplanting period of the summer crop (mid-January to early March); in the following year, 44 percent arrived during the transplanting phase of the winter crop (end of June to early August).

The quantity of rice supplied from the South is especially significant when it is viewed in terms of North Vietnam's production. Thus during 1950-54, when rice output in the areas under French control in the North declined sharply because of war, the supply from the South amounted to as high as 54 percent of the total Northern crop (1950).

Other Items

Other items entering trade from South to North include sugar, tobacco, peanuts, vegetable oils, rubber, and beverages. Cochinchina's large-scale production of refined sugar, which began in the early 1930s, expanded rapidly from 11,776 tons in 1938 to 16,995 in 1940, but then declined to 14,809 tons in 1942.[21] Together with central Vietnam's sugar, South Vietnam's output in 1941 reached 20,372 tons, capable of supplying an important part of the Northern market.[22]

Raw tobacco and cigarettes were also produced mainly in South Vietnam; in 1942 total output reached 5,186 tons, mainly cigarettes, as compared with 4,487 tons in 1940.[23] Cigarette production was adequate for domestic consumption, and the small amount of cigarettes imported from France (12 tons in 1942)[24] was consumed mainly by the European community and the urban population. Most of the domestically produced tobacco came from the four principal factories that were centered in the Saigon-Cholon area.

In addition to sugar and tobacco products, South Vietnam also supplied North Vietnam with vegetable oils. These oils were made from groundnuts, cotton, kapok seeds, sesame, and castor-oil beans. In 1942 areas under groundnut cultivation in South Vietnam (including Annam) amounted to 39,000 hectares as compared with 5,000 hectares in the North. Groundnut production in the same year was 26,000 tons in the South, but only 3,000 tons in the North.[25]

Because of favorable climate and soil conditions in Cochinchina, rubber plantations, which flourished during the colonial period, represented an increasingly important share of total export earnings—rising from 0.3 percent in 1913 to 65.6 percent in 1946 and then going to 48 percent in 1954.[26] As rubber was produced mainly for exports, the amount of its shipment to North Vietnam reached 581 tons in 1948 or nearly 2 percent of total exports; the annual average for other years was 142 tons or about 0.4 percent. Southern rubber shipped to the North was used mainly by the Hanoi rubber factory, which manufactured miscellaneous finished rubber products.

CONCLUSION

Analysis of the structure of commercial relations between the different regions of Vietnam shows that an important part of the goods produced and consumed within each of the regions into which the country is now divided have moved in intraregional trade. Although the North possesses nearly all the country's industrial resources, its less favorable agricultural conditions make it heavily dependent on the South to support its fast-growing population. South Vietnam, on the other hand, is more than capable of satisfying its requirements for foodstuffs and at the same time can produce sizable surpluses for export to the rest of the world.

In 1939 South Vietnam exported approximately 9 percent, and in 1953, 17 percent, of its rice production to the North. The latter in exchange supplied to the South roughly 7 percent of its coal output in 1939 and 9 percent in 1953. These two main commodities represented in each instance an important share of each region's total exports. Thus in the five years that preceded the partition, North Vietnam's shipment of coal to South Vietnam amounted to an average of 37 percent of its yearly exports, while the South's shipment of rice to the North accounted for an average of 36 percent of its annual exports—representing an almost equal percentage on both sides.

TABLE 2.4

South Vietnam's Rice Supply to
North Vietnam in Selected Prepartition Years
(in metric tons)

Year	Paddy	Rice	Broken	Other	Total
		Annam[b]			
1950	207	28,772	2,587	88	31,654
1951	8	40,688	3,862	143	44,701
1952	32	57,852	6,401	469	64,754
1953	19	65,824	4,621	175	70,639
1954	—	95,586	3,719	178	99,483
1953					
First quarter	4	19,043	790	63	19,900
Second	4	13,909	1,386	55	15,354
Third	88	18,853	2,200	26	21,167
Fourth	3	14,019	245	31	14,298
1954					
First quarter	—	14,696	60	32	14,786
Second	—	21,592	360	3	21,955
Third	—	33,640	1,637	94	35,371
Fourth	—	25,658	1,662	49	27,369
		Tonkin[a]			
1950	2,415	52,066	16,390	7,650	78,521
1951	1,284	30,793	1,381	3,260	36,718
1952	—	21,013	4,564	5,615	31,192
1953	1,017	59,277	26,327	13,933	100,554
1954	407	48,251	9,945	8,576	67,179
1953					
First quarter	—	15,999	10,269	3,251	29,519
Second	239	22,785	10,901	5,868	39,793
Third	373	11,821	4,717	2,690	19,601
Fourth	405	8,672	430	2,124	11,631
1954					
First quarter	—	7,322	2,606	2,473	12,401
Second	407	12,504	1,740	1,741	16,392
Third	—	25,436	5,035	2,849	33,320
Fourth	—	2,989	564	1,513	5,066
		Total to North Vietnam[a]			
1950	2,622	80,838	18,977	7,738	110,175
1951	1,292	71,481	5,243	3,403	81,419

1952	32	78,865	10,965	6,084	95,946
1953	1,036	125,101	30,948	14,108	171,193
1954	407	143,834	13,664	8,754	166,659
1953					
First quarter	4	35,042	11,069	3,314	49,429
Second	243	36,694	12,287	5,923	55,147
Third	381	30,674	6,917	2,716	40,688
Fourth	408	22,691	675	2,155	25,929
1954					
First quarter	—	22,018	2,666	2,505	27,189
Second	407	34,096	2,100	1,744	38,347
Third	—	59,076	6,672	2,943	68,691
Fourth	—	28,647	2,226	1,562	32,435

	North Vietnam Rice Production[a][b]			Southern Rice Supply as Percent of Northern Production
Year	Tonkin	Annam	Total	
1950	145,276	—	—	54[c]
1951	457,074	—	—	8[c]
1952	621,752	106,100	727,852	13
1953	393,638	93,903	487,541	35
1954	—	—	—	—
1953				
First quarter	—	—	—	—
Second	158,626	—	—	—
Third	—	—	—	—
Fourth	235,012	—	—	—
1954				
First quarter	—	—	—	—
Second	—	—	—	—
Third	—	—	—	—
Fourth	—	—	—	—

[a]Excluding areas under Communist control.

[b]Though the 1954 agreement divided Annam roughly into two halves, the figures for Annam in this table are included in those for all North Vietnam.

[c]Southern supply to Tonkin as percent of Tonkin's output.

Note: Figures are only partial since they exclude an important quantity of food supplied through uncontrolled trade and border trade.

Sources: Republic of Vietnam, Institute of Statistics, *Nien Giam Thong Ke*, 1949-50, pp. 113-15, 1954-55, pp. 101-05, 148-49; Institute d'Emission des Etats du Cambodge, du Laos, et du Vietnam, *Compte Rendu des Operations*, (Saigon, 1954), p. 74.

In the case of cement, nearly all exports from North Vietnam were directed to the South during this period.

When roads and railways still linked all parts of the country, some 14 percent of North Vietnamese salaried workers moved to the South each year to earn a living. On average, they represented about 33 percent of the total annual exodus of workers departing from North Vietnam to settle elsewhere in the French Union. The Northern immigrants to the South also sent part of their income back home to support their families. They constituted a safety valve to relieve some of the population pressure in the Tonkinese delta.

The effects of the 1954 division of the country and the cutting of the trade between the two economies, and North Vietnam's effort to negotiate for trade resumption, are illustrated in the appendix to this chapter. The postpartition conditions and recovery are analyzed in the next chapter.

APPENDIX: NORTH VIETNAM'S FOOD PROBLEM: PAST EFFORTS TO RESUME TRADE WITH SOUTH VIETNAM*

To North Vietnam, partition meant that beginning in July 1954 it could no longer depend on the South for yearly relief. The food shortage situation at the time of division was further aggravated by the disruption of production and marketing during years of hostilities in the Tonkin Delta and by natural calamities during 1954 and 1955. Only with substantial grain assistance from the Communist countries was widespread famine averted. In 1955 alone, the Soviet Union sent 170,000 tons of rice and China 50,000, making a total of 220,000 tons. It is interesting to note that the latter figure corresponds roughly to the annual flow of paddy from the South in pre-war years.

More than anyone else, North Vietnam's leaders realize the far-reaching implications of the food crisis. Throughout the colonial war, the main theme of the campaign to mobilize public support was centered on the issue of food. Farmers were told that rice was the aim, Communism the means to achieve it. It is natural that when victory over the French came in 1954, the leaders were under great

---

*Excerpted from the author's "Prospects for Economic Cooperation Between the Two Vietnams: Implications for Indochina's Future," in Indochina: Perspectives for Reconciliation, ed. Peter A. Poole (Athens, Oh.: Center for International Studies, Ohio University, 1975).

pressure to fulfill their promises to provide an adequate food supply for the populace. As Bernard Fall pointed out, during the period between the Viet-Minh revolution and the division of the country (1945-54), the Communist Party could still blame any economic crisis on the French expeditionary corps and the war and could turn a crisis into an effective weapon of propaganda. Beginning in 1955, however, the regime has had to assume "the full obligations of orderly government in the economic field. On its eventual success or failure in providing people under its control with an adequate living may well depend in great part the survival of the whole regime."

Fully aware of the vulnerability of Northern agriculture and of the consequences of food deficits, North Vietnam's leaders turned to the South as soon as their administration was established in Hanoi.

The Observer (London), on March 20, 1955, reported a story about Phan Anh, then Hanoi's Minister of Foreign Trade. In an interview, Phan Anh said: "Rice! First and foremost there must be enough rice." He shook his head and made a small decisive motion with his chopsticks. "This country must be unified. They need each other economically, the North and the South. There can be no question of continued partition."

As early as February 4, 1955, seven months after the partition of the country, Hanoi proposed the restoration of "normal relations" including postal exchange and sea, air, and railway transport between the two zones.

Given the shaky state of security in the South during the years following the division of the country and on account of bitter experiences between Nationalists and Communists in the past, it was only natural that the response from President Ngo Dinh Diem was negative.

Furthermore, there was a fear at the time that the trade proposal was only a scheme to permit Northern soldiers to infiltrate the South for subversive purposes.

According to a North Vietnamese publication, Hanoi officially proposed a conference to negotiate economic relations between the two zones on June 6, 1955. The negotiations were to begin on July 20. On July 19, one day before the proposed date of negotiations, North Vietnam sent an official message urging the South to choose "a meeting place on the territory of Vietnam to discuss the matter." According to the same source, South Vietnam turned down the proposal on August 9, 1955.

An incident in Saigon in 1956 provides an important but neglected account of these economic negotiations. In that year, when the food shortage in North Vietnam became critical, Hanoi sent a special mission to Saigon headed by Van Tien Dung, one of the Dien Bien Phu generals and later Chairman of the North Vietnamese Army's Chiefs of Staff; he was to negotiate with President Diem for the resumption of trade. Thousands of students staged anti-trade demonstrations and mounted an attack at the Majestic Hotel in Saigon where the Northern delegation was staying.

Subsequent to the Dung mission, Hanoi re-approached the South on the same subject on March 7, 1958, and on October 4, 1960. According to Jean Lacouture, Ho Chi Minh had expressed his desire on several occasions "to establish commercial relations permitting the North to buy rice from the South at a reasonable price." In retrospect, it seems relevant to point out that it was only in December 1960, perhaps when the hope for trade with the South completely faded, that Hanoi launched the "National Liberation Front" as an instrument for total conquest.

By 1963, when the economic crisis was deepening due in part to serious drought and typhoons between the Spring of 1961 and Winter of 1962, Hanoi reportedly called again for trade between the two zones.

In addition to economic trade, North Vietnam also wanted to negotiate monetary cooperation. In an article entitled, "The Struggle to Establish Monetary Relations between North-South," in Hanoi's Ngan Hang review, the complaint was made that the National Bank of South Vietnam had ignored four letters of the Director-General of the North Vietnam Central Bank proposing the establishment of monetary relations between the two zones; according to this document, the last letter was sent on November 11, 1958.

The primary objective of the proposed monetary cooperation at that time was, however, to facilitate the transfer of funds by Southern Communists who had left the South and gone North after the 1954 cease-fire (that is, those among the founders of the NLF) to their families still remaining in the South.

Whether these proposals were actually made is another question; the fact that the Director-General of North Vietnam's Central Bank took pains to report at length on the issue suggests that there was indeed pressure for such a course of action.

In light of the close economic link between North and South Vietnam in the past and considering the developments within the two zones since partition, it seems realistic to suggest that, as an alternative to the 1965-75 war the two Vietnams could have coexisted under arrangements for a partial unification following the German formula of "Two Zones in One Economic Entity".

That is to say, politically they may remain temporarily as two separate states, but economically they can build a framework of joint economic cooperation eventually leading toward a common market for the well-being of all Vietnam.

That this was a real possibility can be supported by a precedent in Vietnam. In the past there had actually been a precedent involving economic trade between the Communist and non-Communist areas in Vietnam.

Prior to the 1954 partition, the country was already separated into areas controlled by Communists and by non-Communists. At the beginning of the Indochina war, the Communist Party maintained a policy of strict economic embargo vis-a-vis the non-Communist area. Later on, however, realizing the advantages of inter-zonal trade, this policy was completely reversed and trade between the two areas was actually advocated by the Viet-Minh.

By virtue of a decree signed on August 13, 1951, the Communist Party laid down rather classic principles for trade with the non-Communist area, viz., that the trade must (1) secure economic independence for the Viet-Minh zone, (2) protect its production, and (3) induce the import of what was strictly necessary while exporting only what was produced in surplus.

During 1951-52, the level of exchange between the two zones saw a five-fold increase; in 1953, exports from the Communist area to the non-Communist area increased further by 160 percent, while imports increased by forty percent.

An important part of the trade took the form of bartering transactions; the rest involved the use of the currencies of the two zones—the Indochinese Piaster and the Communist Dong. The exchange rates between the two fluctuated widely from one region to another, and from one day to the next. As the Viet-Minh gained strength and their trade position improved, however, the Dong also increased in value. The exchange rate moved from one Piaster to between 48 and 100 Dong (depending on the region) in 1953 to about thirty Dong by the time of partition.

The two segments of population that were involved in warfare and territorial divisions in 1951 and at the same time were also engaged in economic interchange were also the two segments that during 1955-75 largely made up the people of North and South Vietnam, the two parties in the conflict.

Looking back to history and reflecting on the developments in Vietnam in the past decade, one wonders if there was indeed a real possibility that trade could have emerged as a solution leading to real peace in Vietnam and the dramatic outcome of the struggle could be so different from what happened in the spring, 1975.

NOTES

1. The levels of interregional trade in the late 1950s were estimated at 10 percent in Latin America, 4 percent in Central

America, 4 percent in North Africa, and 6 percent in the Arab League. See R. L. Allen, "Integration in Less Developed Areas," Kyklos 14, no. 3 (1960): 321.

2. Estimate is based on the 1954 distribution of population in ten provinces for which data are available. The population under French control was 1,756,000 persons of a total 3,011,000 persons. See RVN, Institute of Statistics, Annuaire Statistique du Vietnam, vol. 5, 1953-54, p. 25.

3. See Charles Robequain, The Economic Development of French Indochina (New York: Oxford University Press, 1944), p. 252.

4. For more details, see Virginia Thompson, French Indochina (New York: Macmillan, 1957), pp. 114-15.

5. RVN, Institute of Statistics, Annuaire Statistique du Vietnam, vol. 1, 1949-50, p. 220.

6. Institut d'Emission de Etats du Cambodge, du Laos, et du Vietnam, Rapport Annual, 1954, p. 122.

7. For figures on import in the 1950s, see U. S. Operations Mission in Vietnam, Annual Statistical Bulletin, 1960, p. 106.

8. Robequain, op. cit., p. 271.

9. RVN, Institute of Statistics, Annuaire Statistique de l'Indochine, vol. 10, 1941-42, pp. 100-01.

10. Institut d'Emission de Etats du Cambodge, du Laos, et du Vietnam, op. cit., 1954, p. 95.

11. RVN, Annuaire Statistique du Vietnam, vol. 1, 1949-50, p. 134.

12. Girard Chaliand, The Peasants of North Viet-Nam (Middlesex, England: Penguin, p. 25.

13. For more details on the contract system, see Thompson, op. cit., pp. 148-54.

14. The Northern peasants who came to the South were often disliked by the Southerners, who were not only suspicious of the migrants, but were also jealous of their reputation as hard and skilled workers. For a thorough analysis of the problem, see J. Gaudal, Problemes du Travail en Indochine; Etudes et Documents (Geneva: Bureau of International Labor, 1937).

15. According to estimates from the French labor inspection of Indochina in 1937, there were 72,000 wage earners, of which 50,000 worked in mines and quarries. Total wage-earning population was expanded to 150,000 in the late 1930s. See Robequain, op. cit., p. 79.

16. Commenting on the immigrant situation, Robequain wrote: "Miserable as was his life there, the man from the north is still homesick for his native village. The discipline of the plantations depresses him. The surroundings, even though more healthful, are

offensive to him. He cannot get used to the forest or the mountains which hem him in on all sides; he longs for the day when he will see them no more. There has been curious proof of this state of mind. When, for example, the trai [camps] were pretty generally replaced with houses, the coolies were extremely reluctant to abandon the collective dwellings which they knew were temporary. It seemed to them that their new one-family homes were traps which would keep them against their will in this land of exile" (ibid., p. 218).

17. For more details on migration, see ibid., pp. 70-72.

18. Ibid., pp. 70 and 217.

19. Nguyen Van Vinh, La Reforme Agraire au Vietnam (Louvain: Universite Catholique de Louvain, 1961), p. 13.

20. Warren S. Thompson, Population and Progress in the Far East (Chicago: University of Chicago Press, 1968), p. 328; and V. A. Zalentsov, "Agrarian Reform in the Democratic Republic of Viet-Nam," Voprosi Economiki Stran Vostoka (Moscow), 1958, pp. 38-70.

21. RVN, Institute of Statistics, Annuaire Statistique de l'Indochine, vol. 12, 1947-48, p. 113. The largest sugar factory owned by the Societe des Sucreries et Raffineries de l'Indochine was built at Hiep-Hoa using sugar cane grown at the eastern Vaico Valley. The sugar-cane plantation represented "an interesting example of collaboration between the European and natives." See Robequain, op. cit., pp. 234-45 and 278.

22. Total Vietnam sugar imports in 1941 amounted to only 240 tons as compared with 1,256 tons in 1939 (RVN, Annuaire Statistique de l'Indochine, vol. 10, 1941-42, p. 158).

23. RVN, Annuaire Statistique de l'Indochine, vol. 12, 1947-48, p. 113.

24. RVN, Annuaire Statistique de l'Indochine, vol. 10, 1941-42, p. 158.

25. Ibid., p. 89.

26. For a detailed history of the introduction of rubber to Indochina, see Thompson, op. cit., pp. 130-36.

PART

II

RECONSTRUCTION AND TRANSFORMATION

CHAPTER

3

PARTITION AND RECONSTRUCTION

Seen against the background of the economic relationship between the two Vietnams as analyzed in Chapter 2, the results of the 1954 partition were inevitable. Each region was cut off from both market and source of supply on which it had previously relied for a large portion of its day-to-day commerce. The effects of the partition were, therefore, damaging to both Vietnams. To North Vietnam, however, the damage appears to have been much more serious because of the nature of its dependence on the South for food.

Furthermore, years of hostility had left visible imprints upon the whole North Vietnamese economy by the time of partition. The gigantic task that the Communist regime set out to undertake had two aspects: physical reconstruction and Communist transformation. The primary objective of economic policy during the immediate post-partition years was not only to adjust to the partition and recover from the war, but also to transform the North Vietnamese economy from mainly private enterprise to a controlled economy strictly modeled after the Chinese and Russian systems. The latter task was carried out in two ways: radical land reform in the agricultural sector and consolidation of state power in the modern sector. In this chapter, only the effects of partition and reconstruction on infrastructure are analyzed. The land reform in the agricultural sector will be discussed in the next chapter.

CONSEQUENCES OF WAR AND PARTITION

At the time of partition, North Vietnam's agriculture was a depressed sector even though it included 12.6 million people, or nearly 93 percent of the population.[1] Disruption of production brought about by over a decade of warfare left the rural sector one of the poorest in Asia. Furthermore, as the rural market became disorganized, the quantity of food marketed also declined significantly. In 1953, for example, the marketed production in 14 delta provinces under French control amounted to only 394, 000 tons, or about 25 percent of peacetime production.[2] In 1955 total food production, although substantially higher, was estimated at just a little over 2 million tons. With a population of 13. 6 million in that year, that output implied a deficit of about 1. 1 million tons.[3] An important part of cultivable land was abandoned during the war years as security conditions in the countryside worsened and also because the irrigation system was damaged by bombings. According to North Vietnamese sources at the time of the cease-fire, 143, 000 hectares of rice fields lay fallow eight of the most productive hydraulic systems capable of irrigating 320, 000 hectares were destroyed, and more than 100, 000 buffaloes and oxen were lost.[4]

The partition also resulted in the departure of nearly 1 million people from the delta seeking refuge in South Vietnam. The exodus was possible because of the provision of the Geneva Peace Agreement (Chapter 1, Article 2) permitting free movement between the two zones for a period of 300 days following the cease-fire. To the extent that most of the refugees were farmers and skilled laborers, their departure meant a serious reduction in North Vietnam's active productive population, which upset economic activities in several areas. In 1955 North Vietnam's population was estimated at 13. 6 million. Assuming that the 1955 structure of population was the same as that given by the 1960 census,[5] the active population in 1955 between the ages of 16 and 55 was 6. 4 million. The exodus of 1 million persons from the North thus implied a reduction of North Vietnam's total population by 7 percent and of its most productive population by 16 percent. According to official sources, North Vietnam suffered a severe shortage of skilled workers at the time. In Hanoi, for example, there were only 5, 000 among a population of 400, 000, or over 1 percent.[6] At the textile plant of Nam Dinh, North Vietnam's largest manufacturing outfit, no engineers were left in 1955. Under normal conditions, the factory required at least 47 engineers to work with 12, 000 workers.[7] The entire operation of the Hon Gai coal mine was under the management of two technicians as compared with 150 engineers previously employed. The industrial sector also lacked capital equipment because an important number of machines and tools in the French-owned factories had been dismantled and transported elsewhere when the French forces withdrew from North Vietnam.

RECONSTRUCTION OF INFRASTRUCTURE

The most pressing task facing a country after any war is that of reconstruction of its physical assets. While relying mainly on institutional reform to recover production in the agricultural sector, Hanoi's leaders concentrated on the reconstruction of the irrigation network, transport facilities, and manufacturing during 1955-57.

The sectoral distribution of public investment during the reconstruction program is presented in Table 3.1. It seems clear that the allocation of financial resources, which reflected the planners' choice and planning strategy during the first reconstruction plan (1955-57), would shed light on the direction and structure of the subsequent reconstruction plan, since Hanoi's economic planners today are the same authors of the first reconstruction plan, and the ideological base of that plan also remains unchanged (see Chapter 10).

In the 1955-57 reconstruction plan, the planners relied heavily on agrarian reform and on the contribution of labor from the rural population to reconstruct the rural sector, especially to repair the irrigation network. The government's efforts in terms of investment were concentrated mainly on transport and communications. Among investments in transportation, repair of rail lines received first priority as the Northern leaders wanted to link their capital to that of China and other socialist countries as well as to connect industrial centers with the new steel and iron mills at Thai Nguyen. Second to transport and communications, budgetary expenditures were allocated for restoration of light industry and manufacturing in order to meet basic consumer needs, which had suffered a critical shortage by the time hostilities had ended.

During the period of reconstruction a total of 559.3 million dong, equivalent to U.S. $158.7 million, was invested in the productive sector, averaging 186.4 million dong per year (U.S. $53 million). In terms of total annual budgetary expenditures, capital spending amounted to 40 percent in 1955, 50 percent in 1956, and 53 percent in 1957.[8]

Irrigation

The vulnerability of North Vietnam's agricultural lands to the caprices of the Red River, as already discussed, necessitates extensive drainage and flood-control facilities in order to stabilize crops. The first priority in rural reconstruction was, therefore, given to restoration of the irrigation system. While major hydraulic works, including the three main dams of Bai Thuong, Thac Huong, and Do Luong were to be repaired by the central government through

TABLE 3.1

Sectoral Distribution of Public Investment under Reconstruction Plan, 1955-57
(in millions of dong)

Sector	1955		1956		1957		Total Reconstruction	
	Amount	Percent	Amount	Percent	Amount	Percent	Amount	Percent
Productive	130.9	93	220.8	85	207.6	83	559.3	86
Agriculture	22.8	16	36.3	14	27.5	11	86.6	13
Irrigation	21.0	15	28.9	11	21.6	9	71.5	11
Industry	9.7	7	79.6	31	105.0	42	194.3	30
Manufacturing	1.4	1	21.5	8	37.1	15	60.0	9
Transport and communications	89.1	63	87.6	34	59.2	24	235.9	36
Railways	65.2	46	40.6	16	19.0	8	124.8	19
Commerce	9.3	7	17.3	7	15.9	6	42.5	7
Nonproductive	9.6	7	39.3	15	43.4	17	92.3	14
Education, culture and health	1.9	1	19.3	7	23.9	9	45.1	7
Research	0.9	1	1.0	—	2.9	1	4.8	1
Housing	1.9	1	5.5	2	4.5	2	11.9	2
Administrative services	4.2	3	11.0	4	9.5	4	24.7	4
Utilities	0.7	1	2.5	1	2.6	1	5.8	1
Total	140.5	100	260.1	100	251.0	100	651.6	100
Equivalent in U.S. $ million	40		74		71		185	

Note: Figures may not add to total because of rounding.

Source: Reconstructed North Vietnamese data, based on DRVN, So Lieu Thong Ke (Statistical data), 1955-59 (Hanoi, 1960), pp. 97-101.

small allocations of investment expenditures, most of the dikes throughout the country were to be reconstructed by the local population.

Irrigation projects involved investment in terms of both money and labor. In addition to the work contributed by the population in the form of investment in labor, most of the budgetary expenditures in agriculture were allocated for hydraulic works.

Total public expenditures in agriculture for the period amounted to 86.6 million dong or 15 percent of total investment. Of this amount, nearly 83 percent (71.5 million dong) went for irrigation. In terms of total economic spending, however, investment in irrigation declined from 16 percent in 1955 to 10 percent in 1957.

With substantial assistance from China, the Bai Thuong dam was restored and the Giang Cao drainage system constructed toward the end of 1957.[9] During 1955-57, six new works capable of providing water for 270,000 hectares were built. In addition, the area irrigated by smaller works reached 368,000 hectares,[10] pushing the total surface area under irrigation to 638,000 hectares, or about 29 percent of total land under paddy cultivation in 1957.

From 1958 to 1960, eight new drainage and pumping stations were built. These works required over 46 million workdays for removing some 37 million cubic meters of earth.[11] As a result of expansion of irrigated land, 45,000 hectares of one-crop land were transformed to two-crop land. By 1955, the total number of hydraulic projects included five major dams, ten drainage systems, and four pumping stations in operation, thus increasing the surface area irrigated to 485,000 hectares (see Table 3.2). Official data for 1960 show total irrigated land, including the area irrigated by large projects as well as by hand, reached over 2 million hectares.[12]

Nevertheless, it seems that these results, taken from official sources, were rather inflated, since they imply a ratio of irrigated area to total cultivable land of 89 percent, a proportion which so far has been reached only by Japan in all of Asia. It seems quite possible, however, that by 1959 North Vietnam had already reconstructed its major dikes and dams and restored the prewar irrigation network. In addition, it built a number of new defensive hydraulic projects to prevent the destruction of crops. In subsequent years, the planners concentrated on building a number of large-scale irrigation networks that would begin to have some positive effect that is, improve the quality of water control and thereby increase crop yields.

One of the advantages of economic reconstruction is that it offers the planners the possibility of making the choice as to what should be reconstructed first and in what way. North Vietnam's planners have opted for water control over and above application of other inputs in order to increase output during the postwar years.

TABLE 3. 2

Irrigation Projects During Reconstruction, 1955-57

	Type of Project		
	Dams	Drainage Systems	Pumping Stations
Number of projects			
Reconstruction	5	6	1
New	--	4	3
Total	5	10	4
Land irrigated by projects (hectares)			
Reconstruction	116. 0	196. 0	14. 5
New	--	134. 0	24. 5
Total	116. 0	330. 0	39. 0

Source: DRVN, So Lieu Thong Ke (Statistical data), 1955-59 (Hanoi, 1960), p. 107.

This choice was a rational one in view of the North's agricultural environment. Furthermore, the production functions of agriculture in Southeast Asia in general also lend support to the view that the best investment at the stage where rice yield is still relatively low would be investment in water control. Table 3. 1 indicates that North Vietnam's investment structure in agriculture in the 1955-57 period review falls within the pattern of investment suggested for the area as a whole.

Transport and Communications

In terms of North Vietnam's investment in financial resources, transport and communications represent the most important items. It was evident that, without an adequate transport system, very little progress could be made in the country's economic development.

During the war years the transport network had suffered the heaviest damage of all sectors. The damages were not only inflicted first by Allied bombing during World War II and later by the French air force, but also by the Communists themselves following their own strategy of "Destroy to resist" (Tieu tho khang chien). Under this strategy destruction of transport and communications was called for in order to isolate the French-controlled area. According to North Vietnamese sources, the Communist insurrectionists deliberately destroyed 10,700 kilometers of roads, 1,540 kilometers of railway track, and 30.5 kilometers of bridges, as well as 53 small towns during the resistance war (1945-54).[13] Major highways linking Saigon with Hanoi, and Hanoi with the Chinese border were also rendered unusable.

Of the 2,597 kilometers of railroad throughout Vietnam, 1,164 kilometers, or 45 percent, were located in North Vietnam. In addition to including a portion of the Trans-Vietnam railway, the rail system embraces three other lines, two connecting Hanoi with the Chinese border, and the third one connecting Hanoi with Haiphong. At the end of hostilities, only the Hanoi-Haiphong railroad (104 kilometers) and a short spur from Hanoi to Van Dien (10 kilometers) were in operation.

During the reconstruction period, top priority was given to the Hanoi-Muc Nam Quan railroad (162 kilometers), which connected Vietnam with China at Nam Quan, and to other lines such as the Hongay-Haiphong-Hanoi route to the East, Hoabinh-Sonla-Dien Bien Phu to the West, Cao Bang-Bac Can-Thai Nguyen to the North, and Ninh Binh-Dong Hoi to the South.[14]

Of total government spending for reconstruction, transport and communications received the highest share, 236 million dong (U.S. $67 million), or 42 percent. Of this total, repair of railroads accounted for more than one-half, 125 million dong (U.S. $35 million). The planners started out to restore the railway system in the very first year of reconstruction and allocated to it 50 percent of total economic investment in that year as compared with 18 percent and 9 percent in the following two years.

In early 1955 the Hanoi-Muc Nam Quan railroad was completely rebuilt with an important amount of economic aid from China, thus linking North Vietnam with China and the Soviet Union. China also provided locomotives, passenger cars, and technical assistance to bring the Hanoi-Nam Dinh line back into operation toward the end of the same year, thereby connecting the capital with the country's biggest textile center. From Nam Dinh, the the repair work continued southward to Ninh Binh and to Thanh Hoa. A second line linking the capital with China, the Hanoi-Lao Cay route, was also completed in 1956.[15]

By 1957 a major part of the transport network had already been repaired. The length of the main railroad in operation reached 57 percent of the prewar length, while that of primary roads (about 3,310 kilometers) was also reconstructed.

TABLE 3. 3

Share of Transport Modes in
Total Merchandise Transport, 1955-60
(in percentage)

Year	Rail	Road	River	Ocean
1955	14	23	62	1
1956	19	31	48	1
1957	19	41	39	1
1958	20	39	38	2
1959	21	38	39	2
1960	23	35	40	2

Note: Figures may not add to 100 percent because of rounding.
Source: So Lieu Thong Ke (Statistical data), 1955-59, (Hanoi, 1960), p. 177.

In addition to roads and rail, North Vietnam recovered river navigation, a large amount of which was also disrupted during the war years. Traditionally, waterways provided a very important means of transportation for the Tonkin delta; the length of the main navigable rivers reached over 4,000 kilometers in 1959.

In 1955, river transport accounted for 62 percent of merchandise transported compared with 14 percent by rail and 23 percent by road. As reconstruction proceeded, however, the transport pattern changed decidedly in favor of roads and rails (see Table 3. 3).

Beginning in 1965, the changing trend was reversed as North Vietnam again resorted to rivers for internal transport under the impact of U. S. bombing.

During the reconstruction program, ports were also reconstructed for international trade. The five major ports of Haiphong, Hongai, Campha, Cua Ong, and Ben Thuy were enlarged. The Haiphong harbor, largest in Southeast Asia, is situated 60 miles southeast of Hanoi on an arm of the Thai Binh River, which flows into the Gulf of Tonkin. Haiphong is the principal port of North Vietnam and as such is of crucial importance to the economy. In 1955 it was not capable of operating at full capacity because of the lack of adequate dredging during the war years. As a consequence, large vessels had to transfer their loads to smaller barges in the outer bay. Annual dredging, which was not done satisfactorily in 1953-54, is required to enable large oceangoing ships to reach the port, which is situated ten miles from the sea. With the supply from the Soviet Union of a large harbor dredge, however, Haiphong was capable of receiving 10,000 ships annually by 1957.[16]

Light Industry

According to official data, industrial production in North Vietnam at the time of the cease-fire in 1954 amounted to only 1.5 percent of GNP as compared with about 10 percent in prewar times.[17] The pressure on the market for manufactured goods was further aggravated by the sudden discontinuation of French imports. As a consequence, a serious shortage of consumer goods, such as textiles, soap, paper, sugar, matches, and chemical products, was made even more acute.[18] Although Hanoi's planners had already decided to favor development of heavy industry, they were also inclined to pay some attention to the manufacturing sector during 1955-57 in order to alleviate the critical shortages in essential consumer goods. Thus, as soon as it settled in Hanoi in September 1954, the Politburo issued a policy directive related to priority in reconstruction:

> At this time it would be impossible to implement a plan to construct heavy industry and to do so with fast progress. Our need for the time being is to reconstruct and to build a number of plants producing basic goods for the public, and shops to repair transport equipment. These undertakings, while requiring little capital for quick return, may solve pressing problems for the people.[19]

It was also emphasized that manufacturing plants could be built quickly and could thus immediately absorb the unemployed, unskilled workers in the urban areas, estimated at 140,000 in 1955.

Among projects on top of the list for construction were the Nam Dinh textile and Haiphong cement complexes, the Phu Tho tea plant, the Thang Long cigarette plant, the Thong Nhat match factory, the Haiphong shipyard, the Ha Long canned-fish factory, the Gia Lam railroad workshop, and the Hanoi machinery plant. Aid from the Communist countries was quick to come and became one of the major criteria in North Vietnam's choices of project investments among alternatives.[20] Most important aid came from China, which, together with Czechoslovakia, reconstructed and supplied the Nam Dinh textile plants. The Soviet Union helped to restore the Haiphong cement factory to full capacity and to build the Hanoi machinery plant.

From 1955 to 1959, investment in industry amounted to 194.3 million dong, or 35 percent of total economic spending. Most of the investment was made in 1956 and 1957 rather than in 1955 as was the case in railroad reconstruction.

North Vietnam's planners began reconstruction and development of the industrial sector by classifying it into two subsectors: group A, which included public utilities and heavy industry; and group B, light industry and manufacturing. The amount of resources allocated for manufacturing and light industry (group B) was 60 million dong over three years, averaging 20 million dong, or 11 percent of the total industrial investment per annum.

Investment in group A over the three-year period was 134.3 million dong, or 24 percent of the total, mainly for reconstruction of electric-power plants and the Haiphong cement plant, and for construction of the new Hanoi machinery plant. On a yearly basis, the allocation of investment funds within the industrial sector reflects the planners' increasing interest in the manufacture of consumer goods under group B during the period of reconstruction. The share of group B increased from 1.4 million dong, or 15 percent of the amount allocated for industry (11 million dong), in 1955 to 37 million dong, or 36 percent, in 1957. At the same time group A investment declined from 85 to 65 percent in those three years.

Results of reconstruction claimed by the Hanoi leaders were impressive. According to official information, the volume of industrial output in 1957 had already reached the 1939 level.[21] From 317 million dong in 1955, the value of industrial output rose to 865 million dong in 1957—that is, it more than doubled in three years. Of these totals, the rate of increase in production of consumer goods (group B) was higher than that of capital goods (group A). From a total of 231 million dong, or 73 percent of the value of industrial output in 1955, the value of manufactured products increased to 661 million dong, or 77 percent of the total in 1957 (see Table 3.4). The speed of expansion of consumer-goods production within the state-enterprise sector grew at an even faster rate. The value of state enterprise in group B, only 5.6 million dong, or 15 percent of total output of the public sector in 1955, increased to 155.6 million dong, or 63 percent in 1957.

CONCLUSION

While the development and performance of the industrial sector will be analyzed in detail later in this book (Chapter 8), it is sufficient to conclude here that, during the period of reconstruction, North Vietnam recorded significant progress, perhaps unparalleled in its entire history of economic development thus far. It seems, therefore,

TABLE 3.4

Value of Industrial Output During Reconstruction, 1955-57
(in millions of dong)

Industry Sector	1955		1956		1957		Total Reconstruction	
	Amount	Percent	Amount	Percent	Amount	Percent	Amount	Percent
Group A—public utilities and heavy industry	85.8	27	180.1	29	203.9	24	469.8	26
Group B—light industry and manufacturing	230.8	73	446.8	71	661.0	76	1,338.6	74
Total	316.6	100	629.9	100	864.9	100	1,811.4	100
State enterprise								
Group A	32.6	85	83.1	53	90.2	37	205.9	47
Group B	5.6	15	74.2	47	155.6	63	235.4	53
Total	38.2	100	157.3	100	245.8	100	441.3	100

Source: DRVN, So Lieu Thong Ke, (Statistical data), 1955-59 (Hanoi, 1960), pp. 154, 156.

that there is some validity to North Vietnam's frequent claim that by the end of 1957 the task of reconstruction had been largely completed in major sectors.

Reconstruction priority was given to infrastructure, especially in the rebuilding of the irrigation and railroad networks. While the path toward building a strong heavy industry had been considered by the Communist leaders at the very beginning of their administration as the prime target for their country, manufacturing and light industry received a great deal of attention during the immediate postwar years. Limited progress was recorded in several industrial projects, particularly the Hongai mining complex, the Tran Hung Dao machinery plant in Hanoi, the plywood factory at Cau Duong, and a number of electric-power plants.[22]

One main feature of the reconstruction period was the speed at which North Vietnam's economy underwent structural and institutional changes. Side by side with economic reconstruction, the government carried out its second task of transforming the economy from private ownership to collective and from private enterprise to a state-controlled system. The transformation was considered by the Hanoi leaders as essential in view of their ideological pursuit and their plan to industrialize the economy following the Marxist-Leninist model. The next chapter examines the main features of the structural transformation.

NOTES

1. DRVN, So Lieu Thong Ke (Statistical data), 1955-59 (Hanoi, 1960), p. 71.

2. Of the total, 159,000 tons were from the fifth-month crop and 235,000 from the tenth-month crop. See RVN, Annuaire Statistique du Vietnam, 1953-54, p. 104.

3. DRVN, So Lieu Thong Ke, 1955-59, op. cit.

4. D. T. Truyen and P. T. Vinh, Building an Independent National Economy in Vietnam (Hanoi: Foreign Languages Publishing House, 1964), pp. 45-46.

5. The 1960 census gave 47 percent as the size of the population between ages 16 and 55. See DRVN, So Lieu Thong Ke, 1955-59, op. cit., p. 64.

6. Truyen and Vinh, op. cit., p. 46.

7. Vo Nhan Tri, Croissance Economique de la Republique Democratique du Vietnam (1945-65) (Hanoi: Editions En Langues Etrangeres, 1967), p. 228.

8. DRVN, So Lieu Thong Ke, 1955-59, op. cit., p. 77.

9. DRVN, Economic Institute, Kinh Te Vietnam, 1945-1960 (The Vietnamese economy, 1945-1960) (Hanoi: Su That Publisher, 1960), p. 240.

10. DRVN, Etudes Vietnamiennes (English edition) no. 13 (Hanoi, 1967), p. 30.

11. Ibid.

12. Of the 2,024,000 hectares of irrigated land given for 1960, 932,000 hectares or 46 percent belonged to the fifth-month crop, and 1,092,000 hectares, 54 percent, to the tenth-month crop (ibid., p. 30).

13. Truyen and Vinh, op. cit., p. 30.

14. Vo Nhan Tri, op. cit., p. 232.

15. Over this line Chinese tin from Kokiu mines in Yunnam province was shipped to Haiphong before World War II. The Chinese rebuilt their part of the railroad, which was damaged during the war, in order to resume traffic between their province and North Vietnam's port of Haiphong. See Theodore Shabad, "Economic Development in North Vietnam," Pacific Affairs 31, no. 1, (March 1958): 49.

16. Haiphong harbor was originally designed to handle 1.6 million tons of goods a year. In 1955 the volume handled was 500,000 tons, or less than one-third of the normal load (ibid., p. 50).

17. DRVN, Kinh Te Vietnam, 1945-1960, op. cit., p. 237.

18. As a result of shortages in consumer goods, prices soared and in some cases tripled and quadrupled, or even increased tenfold during 1956-57 (ibid., pp. 240-41).

19. Ibid., p. 176, "Nghi Quyet Cua Bo Chinh Tri, thang 9, 1954" (Resolution of the Politburo, September 1954).

20. Shabad, op. cit., p. 39.

21. DRVN, So Lieu Thong Ke, op. cit., p. 148.

22. DRVN, Kinh Te Vietnam, 1945-1960, op. cit., p. 179.

CHAPTER

4

SOCIALIST TRANSFORMATION

At the fourth session of the North Vietnamese National Assembly in March 1955, it was decided: "From the economic viewpoint, land reform creates the fundamental conditions to restore the economy, develop agriculture, restore and develop the urban economy . . . . It will improve the living conditions of the population, especially the peasants."[1] The theme of agrarian reform as the main factor in the restoration of rural productive capacity was also emphasized by the executive branch of the government and by Communist party leaders during the same period.[2] To Hanoi's leaders, land reform was viewed as not only an essential political objective, but also as an effective way to mobilize the masses and to bring a large amount of land abandoned during the war years back into production. By increasing ownership (through redistribution of land) and lowering production cost (through reduction of rent), the Communist leaders expected the peasants to increase food production because of new personal incentives. With the slogan, "He who tills the land will keep it for himself,"[3] the peasant was made to believe that he now owned a plot of land, and eventually he would work on it harder.

However, land reform was only the first step in the new government's strategy of communizing the rural sector. The plots of land that the peasants had received were soon taken away through the process of collectivization.

In the modern sector, the strategy of communizing the economy was carried out by extending state control over all major activities and by rapid growth of public enterprises.

LAND REFORM

The agrarian-reform program in North Vietnam began as early as November 1945. Shortly after the August 1945 Communist revolution, agrarian reform was launched with the proclamation of an immediate reduction of 25 percent in all land rents.[4] In spite of the emphasis on rent, however, the main efforts during the period from 1945 to 1953 were actually focused on the redistribution of land confiscated from the French and Vietnamese landlords.

From statistical data covering 3,035 villages, total land reallocated from 1945 to 1957, when land reform ended, amounted to 518,710 hectares (see Table 4.1). The redistribution was implemented in three periods. From 1945 to 1949, 113,306 hectares of land, or 22 percent of (French and Vietnamese) landlord-owned and public properties, were confiscated. During 1949-53, another 189,434 hectares, or 36 percent, were added, making a total of 302,739 hectares. According to official data, the latter figure represented redistribution of 89 percent of all land previously owned by the French and 85 percent of that owned by the Vietnamese landlords.

A more systematic and comprehensive land-reform program was launched in 1953, officially labeled the Land Rend Reduction Campaign. This preceded the Land Reform Campaign proper, which took place from 1954 through 1956.[5] During the latter campaign, 142,323 more hectares were confiscated, representing 27 percent of all land belonging to landlords, churches, foreigners. It is interesting to note that, prior to the Land Reform Campaign proper (1954-56), 73 percent of land belonging to the nonpeasant class had already been seized; the Land Reform Campaign thus affected only the remaining 27 percent. From this observation, one may postulate that the redistribution of land was actually not the primary objective of the 1954-56 campaign.[6]

During the course of implementation of land reform, the population was classified into five categories: landlords, rich peasants, middle-class peasants, small peasants, and landless peasants.[7] Property belonging to landlords was reduced, while that belonging to small and landless peasants was increased.[8] Harsh and cruel measures were applied in the people's courts to condemn landlords, and a large number of people were victims of liquidation.[9] The cruelty of reprisal was so great that the party itself was obliged to take action to correct what it called "grave errors" by carrying out "mistake correction" campaigns.

TABLE 4.1

Land Confiscated, 1945-57
(area in 1,000 hectares;
percentage in relation to total area of each category)

Owner of Confiscated Land	1945-49		1949-53		1953-56[a]		Total 1945-57	
	Area	Percent	Area	Percent	Area	Percent	Area	Percent
French citizens	11,656	61	5,324	28	2,054	11	19,034	100
Vietnamese landlords	25,522	10	75,366	31	143,562	59	244,450	100
Religious properties	—	—	—	—	15,411	100	15,411	100
Public and semipublic domain	76,127	32	108,744	45	54,943	23	239,815	100
Total	113,305	22	189,434	37	215,970	42	518,710[b]	100

[a]Including land seized during 1953 and during the Land Reform Campaign (end of 1953 to 1956).

[b]This amount is substantially lower than total results of land redistributed up to 1957, officially given as 810,000 hectares.

Note: Results were for 3,035 villages in North Vietnam; individual items may not add to total because of rounding.

Source: DRVN, Economic Institute, Kinh Te Vietnam, 1945-60 (Hanoi: Su That Publisher, 1960), pp. 66, 83.

On a national basis, official data on the results of rural reform show that, by the end of 1957, 810,000 hectares, or about 37 percent of cultivable land, were distributed to 2,104,000 families (9.5 million peasants), implying a figure of 0.4 hectares of land per family.

COLLECTIVIZATION

In strict application of the Chinese experiment in transforming the rural sector, North Vietnam's leaders introduced collectivization in the countryside after land reform. The process of collectivization included three phases: a loose form of cooperation by work-exchange teams (or mutual-aid teams); the establishment of low-rank cooperatives; and the advancement of cooperatives from low-rank to high rank.[10]

Under the work-exchange team arrangement, groups of farmers pooled their resources, especially labor and equipment, in order to undertake joint agricultural projects. These teams could be either seasonal or permanent in nature. Members of seasonal teams undertook collective work during certain times of the year such as the planting or harvesting periods of rice cultivation. The permanent team, on the other hand, was a loose arrangement through which farmers continued to work on their own land while giving a helping hand to others when called upon and on a reciprocal basis.*

In 1955 there were 1,094,000 families, representing 41 percent of the total farm population, who joined 153,000 work-exchange teams. However, the number of members as well as the number of work-exchange teams declined sharply in 1957 as the peasants reacted strongly to the party's harsh land-reform measures and attempted to return to private undertakings. The number of work-exchange teams, which had increased from 153,000 units in 1955 to 190,000 units in 1956, declined to 101,000 units (or by 47 percent) in 1957, and the number of families joining these teams declined from 1,352,000 in 1956 to 595,000 in 1957 (or by 56 percent). Following the decline, however, countermeasures were taken by the government, and the number of family members rose again, to 1,037,000, representing 37.5 percent of the total farm population in 1959 (see Table 4.2). However, most of the teams were seasonal rather than permanent.

The work-exchange teams were only "the seeds of communism," through which the peasants were introduced to the ultimate phase of

---

*The permanent teams were also called year-round teams. In joining a permanent team, a member was paid according to workdays and work points.

TABLE 4.2

Collectivization in Agriculture, 1955-72
(units in 1,000; members in terms of farm families joining;
percentage in relation to total farm families)

Year	Work-Exchange Teams			Agricultural Cooperatives		
	Seasonal	Permanent	Total	Low Rank	High Rank	Total
1955						
Units	141	12	153	7	1	8
Members	1,006	88	1,094	96	10	106
Percentage	38	3	41	—[a]	—[a]	—[a]
1956						
Units	170	20	190	36	1	37
Members	1,216	136	1,352	531	7	538
Percentage	45	5	50	—[a]	—[a]	—[a]
1957						
Units	92	9	101	42	3	45
Members	541	54	595	639	105	744
Percentage	20	2	22	—[a]	—[a]	—[a]
1958						
Units	169	75	244	—	—	—
Members	1,207	570	1,777	126	1	127
Percentage	45	21	66	5	—[a]	5
1959						
Units	65	33	98	27	2	29
Members	725	312	1,037	1,175	66	1,241
Percentage	21	17	38	43	2	45
1960						
Units	—	—	—	37	4	41
Members	—	—	—	2,074	331	2,405
Percentage	—	—	—	74	12	86
1962						
Units	—	—	—	20	10	30
Members	—	—	—	1,430	1,088	2,518
Percentage	—	—	—	49	37	86
1963						
Units	—	—	—	20	11	31
Members	—	—	—	1,355	1,168	2,523
Percentage	—	—	—	46	40	85
1970						
Units	—	—	—	4	18	22
Members	—	—	—	—	—	—
Percentage	—	—	—	20	75	95
1972						
Units	—	—	—	3	17	20
Percentage	—	—	—	—	—	96

[a]Less than 0.05 percent.

Note: Vietnamese term for work-exchange teams is to doi cong. Also called labor-exchange teams or mutual-aid teams, they almost ceased to exist by the end of 1960 and were being transformed into cooperatives. Vietnamese term for agricultural cooperatives is hop tac xa san xuat nong nghiep. Systematically and officially, the cooperative movement began in 1958 with the three-year plan.

Sources: DRVN, So Lieu Thong Ke, 1955-59 (Hanoi, 1960), Tables 23-25, pp. 84-85, and 1960-63, Table 1, p. 39, DRVN, Etudes Vietnamiennes, no. 44, 1976, p. 167.

collectivization. Although begun in 1955, collectivization was only declared as official policy in 1959 when the cooperative movement was formally established. An agricultural cooperative was an arrangement under which members pooled their land and tools together to work under a unified management.[11] In a low-rank cooperative, a member was allowed to receive part of the crop as a form of land payment proportional to the amount of land contributed by him. In a high-rank cooperative, land payment was eliminated[12] and each member was paid solely according to the work he contributed.

Since collectivization was the main feature of Communist transformation in the rural sector, it was carried out with utmost vigor by the party's cadres. From 8,000 units in 1955, the number of cooperatives nearly quadrupled to 28,800 by 1959, embracing 1,241,400 families or nearly half of the farm population. By 1960 the first phase of the cooperative movement was nearly completed as 2,404,800 peasant families, or 85.8 percent of rural families, joined 41,400 cooperatives. However, nearly 90 percent of these cooperatives were still classified as low rank.

As most farmers had already joined cooperatives by 1960, progress in collectivization after that year was made in only two areas: the shifting of cooperatives from low rank to high rank and the consolidation of small cooperatives into bigger ones. Thus, in 1963 the total number of families joining all cooperatives increased to 2,523,100, or by only 5 percent over the 1960 level, while membership in high-rank cooperatives nearly tripled. Several small cooperatives merged into big ones, and the number of ccoperatives declined from 41,400 units in 1960 to 30,600 units in 1963. By 1971 cooperatives were consolidated into 20,000 units covering nearly 96 percent of all peasant families. Most of the cooperatives were also considered to be high rank by that time. Selected features of management of high-rank cooperatives are presented in the appendix to this chapter.

Economic Effect of Institutional Reform

The demoralizing effect of the agrarian reform on the agricultural sector has been the subject of analysis elsewhere.[13] The peasants' resentment of the drastic measures taken by Communist cadres during the campaign is best illustrated by the 1956 open revolt of the peasants from Nghe An province, which was Ho Chi Minh's birthplace.[14] The Nghe An revolt is known only as an isolated incident to the outside world. There is reason to believe that the food crisis was extremely serious and widespread. The graveness of the situation can be measured by Ho Chi Minh's decision to remove Truong Chinh, the most ardent executer of land reform, from his position as the party's secretary general, and by the government's open admission of failures and errors.

Since the Northern regime counted heavily on land reform to boost production, failure of the campaign implied a deepening of the food crisis at the time. By itself, land reform cannot assure higher output, since subdivision of land may even bring temporary or even permanent loss of output per acre. Furthermore, the incentive effect that was supposed to come from increased peasant ownership during the land reform was not at all visible.

On the contrary, lack of interest in collective agricultural activities was evidenced by the fact that there was a sharp drop in the number of farmers joining work-exchange teams during the 1955-57 period, as already noted.

While the magnitude of the agricultural problem will be analyzed later in this book (Chapter 7), it is sufficient to point out here that one may challenge the official claim that by 1955 (the first year of peace), agricultural production had surpassed the prewar (1939) level and that the food-deficit problem had been corrected.[15] It appears from all available evidence that North Vietnam has remained seriously deficient in rice production, and to this day the population still has to resort to inferior staples and imports.

Shortage of food has resulted in strict rationing and price controls over the years. The government has made strenuous efforts to increase production of subsidiary crops, especially maize, sweet potatoes, and manioc. In 1955 areas under cultivation in maize were expanded to 154,000 hectares as compared with 68,000 hectares in prewar times. By 1957 total output of subsidiary crops reached 944,000 tons, or about one-fourth of the rice output.

For purposes of rationing, taxation, and valuation, data for subsidiary crops were converted to paddy data at the ratio of one kilogram of paddy equivalent to 0.7 kilograms of maize, 2.3 kilograms of sweet potatoes, 2.3 kilograms of manioc (cassava), or 0.5 kilograms of soybeans.[16] On the basis of this ratio, the 1957 output of subsidiary crops was valued at 637,000 tons, capable of substituting for an important part of the rice deficit. Nevertheless, consumer taste discriminates grossly against these much inferior substitutes, and people only resort to them in case they cannot obtain rice either through official or black-market channels.

In the face of acute shortages in the food supply, North Vietnam has had to rely greatly on imports. Food imports in the 1950s included sizable aid from Communist countries. In 1955 alone, the Soviet Union supplied 170,000 tons of rice and the Democratic Republic of China, 50,000 tons, a total of 220,000 tons, or roughly equivalent to the annual flow of paddy to North Vietnam from South Vietnam prior to partition.

In 1957 food imports (valued at U. S. $32.4 million) accounted for nearly 33 percent of total imports. The food component, although still exceptionally high, was reduced to 13 percent in 1960 before it rose again to 17 percent in 1963.

It appears that massive external aid and imports, plus the increase in production of subsidiary crops, were the main factors helping to avert mass starvation in the North during the first years after partition.

CONSOLIDATION OF STATE POWER IN THE MODERN SECTOR

Side by side with drastic land reform in the rural sector, consolidation of state power in the modern sector was equally important to North Vietnam's leaders during the period of reconstruction.

The 1954 peace settlement offered the first chance to put into action the Marxist-Leninist theory, which the leaders had long studied. Consequently, the industry and trade sectors were to be transformed into so-called state capitalism, while the handicraft and retail-trade sectors were to be transformed into cooperatives.

The question arose as to how fast and in what way the process of economic transformation would proceed. Through a thorough analysis of economic conditions in the country,[17] the Northern leaders came to the conclusion that the Communist doctrine was to be best pursued "in light of existing economic conditions" of Vietnam at the time by first adopting the so-called peaceful transformation of the economy.[18] According to this approach, the first phase of state capitalism (or state capitalism in its lower form) calls for the state to "redeem rather than to seize immediately machines and equipment from the private sector.[19] In theory, the former owners received payments, in installments over a period of time, amounting to either partial or total value of their assets, plus interest ranging from 4 to 6 percent per year. Previous owners were also guaranteed employment in the reconverted enterprises.[20]

In addition to taking over the public-utilities sector, the Communist regime purchased from the French a number of important plants. Among the companies sold to the government were the coal mines at Hongay, the Haiphong cement plant, the Nam Dinh textile factory, the Hanoi distillery, and the Haiphong phosphate plant. In principle, the coal mines were to be paid in kind, the French receiving a promise to be paid with 1 million tons of coal in 15 years.

Regarding the private sector in general, Hanoi's leaders were rather moderate at first, proceeding only to utilize, restrict, and control the capitalist sector during the first two years after partition. Anxious to recover from the wartime economy, the government at first wanted to rely somewhat on private incentives, the principle being "to utilize private capital and management skills, while restricting the growth of the private sector and controlling speculative activities."[21]

In refraining from complete abolition of the private sector during the reconstruction period, Hanoi appeared to follow the economic policy of the Soviet Union in the early 1920s and based its thinking on Lenin's treatment of the theory of state capitalism as the first step of Communist transformation. In the Soviet Union, when the task of reconstruction loomed large, Lenin made a tactical retreat in 1921 by playing down the need to advance international communism in favor of a new economic policy (NEP), and adopted a policy of moderation toward private ownership and the market economy.[22] The NEP gave strong stimulus to economic expansion, and by 1927 industrial output attained the prewar level while agricultural production reached it a year earlier.[23] The policy of moderate state capitalism adopted in North Vietnam did in fact bring about some results in the earlier days of reconstruction. Progress in the private sector was noted particularly in farm tools, textiles, and timber.[24]

However, by 1957, Hanoi's leaders had evidently grown impatient with the speed at which the socialist transformation of North Vietnam's economy was taking place. Thus, just when some parts of the private sector had begun to flourish, Hanoi applied a brake to the expansion, deciding to take over the private sector. Its decision was further strengthened by its efforts in 1958 to minimize the impact of the "hundred flowers" campaign, a liberal movement taking place among the educated class at the time. Apparently encouraged by the liberal interlude that was going on in China (between October 1956 and June 1957), the Vietnamese intelligentsia had launched the "hundred flowers" campaign, a sort of liberal interlude of their own, in 1958.[25]

The government's policy from that time on was to limit and control private enterprise rather than to utilize it as was the case in previous years. A party directive in 1957 stipulated, "In order to utilize the positive aspect of private enterprise, it is essential to limit its negative aspect." Subsequently, a campaign was launched to condemn profits and tax evasion in the private sector and to accuse it of responsibility for the crisis in the supply of consumer goods during 1956-57.[26] At the same time, the government moved toward completely taking over industry. Progress in consolidation of state power in the modern sector is shown in Table 4. 3.

In industry, the policy of moderation followed during 1955-56 had helped to push the value of private output from 53 million dong in 1955 to 103 million dong in 1956—an increase of more than double

TABLE 4. 3

Growth of State Enterprise in the Modern Sector, 1955-63
(percentage share of total output)

Sector	1955	1956	1957	1958	1959	1960	1962	1963	1965	1968	1971
Industry and handicrafts											
State enterprise	12	25	29	36	48	51	60	61	64	63	64
Mixed state-private enterprise	—	—	—	—	3	6	6	6			
Private enterprise	17	16	14	11	4	—	—	—	36	37	36
Handicrafts (mainly cooperatives)	71	59	57	52	46	43	34	33			
Total	100	100	100	100	100	100	100	100	100	100	100
Trade											
Wholesale											
State enterprise	28	45	53	74	89	94	81	82			
Private enterprise	72	55	47	26	11	6	19	18	—	—	—
Total	100	100	100	100	100	100	100	100	—	—	—
Retail											
State enterprise	20	29	25	33	41	51	82	87	80	76	32
Buying-selling cooperatives	—	4	6	12	21	23					
Autonomous state stores	—	3	9	13	18	17	11	5			
Private enterprise	80	64	60	42	20	9	7	8	20	24	68
Total	100	100	100	100	100	100	100	100	100	100	100
Transport											
Freight											
State enterprise	24	43	50	65	70	75	77	79	41	32	55
Private enterprise	76	57	50	35	30	25	23	21	59	68	45
Total	100	100	100	100	100	100	100	100	100	100	100
Passenger											
State enterprise	49	47	52	—	—	70	74	75			
Private enterprise	51	53	48	—	—	30	26	25	—	—	—
Total	100	100	100	100	100	100	100	100	—	—	—

Note: Individual items may not add up to total because of rounding. State enterprises are under the control of the Ministry of Internal Commerce. In 1962, buying-selling cooperatives were merged into state enterprises.

Sources: DRVN, Economic Institute, Kinh Te Vietnam (Hanoi: Su That Publisher, 1960), p. 212; DRVN, So Lieu Thong Ke (Statistical data), 1955-59 (Hanoi, 1960), pp. 158-60, 179, 190, 193, and So Lieu Thong Ke, 1960-63 (Hanoi, 1963), pp. 44, 91, 97; DRVN, La Republique Democratique du Vietnam (Hanoi: Editions En Langues Etrangeres, 1975), p. 65.

in two years (see Table 4.4). In 1957, however, private output rose by only 19 percent, to 123 million dong. By 1959, private industrial production shrunk to half the 1957 level, and by 1960 it ceased to exist.

In the public sector, the value of the state-enterprise output had more than quadrupled in 1956 to 157 million dong. It then rose steadily to 1,378 million dong in 1963, accounting for 61 percent of the total value of industrial output. In addition to the state enterprises, the mixed state-private enterprises began operations in 1959, with a modest output amounting to 33 million dong. Thus, by 1960, the public sector (including the state and the mixed enterprises) had on the whole replaced private enterprises in industry.

In handicrafts, the Communist transformation took place through the creation of producers' cooperatives. When hostilities had ended in 1954, handicrafts were most important in the industrial-handicraft sector, producing essential consumer goods such as pottery, baskets, textiles, furniture and farm equipment (hand tools, carts, plows), and products for export (lacquerware, carved ivory, embroidered textiles). Handicrafts accounted for over 71 percent of output value in the manufacturing sector in 1955, but this figure began to decline as industrial output in the public sector rose during the following years. Most of the goods produced by cooperatives were sold to the state through the state trading companies. The latter in turn supplied cooperatives with raw materials and tools.

The process of consolidation of state power was also rapid in the area of trade and transport. The task of reconstruction of the transport system, as well as resumption of internal trade that had been disrupted during the war years, offered the possibility of shaping policies for organization and control of the entire distribution and marketing system. In order to restore the level of internal trade, the state trading companies, established before 1954, were strengthened while new ones were added. The government also established state transport companies in order to take over completely the control of distribution. Wholesale trade was gradually taken over by the state trading companies such as the Foodstuffs Company and the General Commodities Company.

At the retail level, transactions were handled by three types of enterprises: state stores, buying-selling cooperatives, and private traders. The state trading companies played an important role, not only as buyers of final products, but also as suppliers of raw materials and other inputs. The function of the state stores was to carry out retail sales to both the buying-selling cooperatives and to individual consumers. The role of private merchants, though important in 1955, declined sharply in the following years, and

TABLE 4.4

Value of Production in Industry and Trade, by Public and Private Sector, 1955-63
(in millions of dong)

Sector	1955	1956	1957	1958	1959	1960	1962	1963
Industry and handicrafts								
State enterprise	38	157	246	372	623	754	1,251	1,378
Mixed state-private enterprise	--	--	--	--	33	87	122	136
Private enterprise	53	103	123	114	59	--	--	--
Handicrafts (mainly cooperatives)	225	366	496	527	596	630	729	747
Total	316	626	865	1,013	1,311	1,471	2,102	2,261
Trade								
Wholesale								
State enterprise	102	227	314	428	620	829[a]	--	--
Private enterprise	260	281	282	154	76	57[a]	--	--
Total	362	508	596	582	696	886	--	--
Retail Sales								
State enterprise	121	274	267	349	514	819	1,501[b]	1,689[b]
Buying-selling cooperatives	2	37	57	122	261	370		
Autonomous state stores	1	30	92	135	223	273	202[b]	97[b]
Private enterprise	485	604	621	447	242	145	128[b]	155[b]
Total	609	945	1,037	1,053	1,240	1,607	1,831	1,941

[a]Preliminary results.

[b]Estimates based on total and percentage shares of each sector (see Table 4.3).

Note: Individual items may not add up to total because of rounding. State enterprises are under the control of the Ministry of Internal Commerce. In 1962, buying-selling cooperatives were merged into state enterprises.

Sources: DRVN, Economic Institute, Kinh Te Vietnam (Hanoi: Su That Publisher, 1960), p. 212; DRVN, So Lieu Thong Ke (Statistical data), 1955-59 (Hanoi, 1960), pp. 158-60, 179, 190, 193, and So Lieu Thong Ke, 1960-63 (Hanoi, 1963), pp. 44, 91, 97; DRVN, La Republique Democratique du Vietnam (Hanoi: Editions En Langues Etrangeres, 1975), p. 65.

was gradually reduced to that of sales agent for the state enterprises.[27]

For example, at the end of 1957, private merchants accounted for 47 percent of wholesale trade and 60 percent of retail trade, as compared with 72 and 80 percent, respectively, in 1955. On the other hand, the state trading companies increased their share in wholesale trade and retail trade to 53 and 25 percent of the respective totals in 1957.

The buying and selling cooperatives and the state stores also grew manyfold. The number of stores run by buying and selling cooperatives increased from 133 units in 1955 to 875 units in 1957, and the number of state stores increased from 474 to 906. This pushed their combined retail sales value from a negligible level in 1955 to 15 percent of total retail sales in 1957, and 40 percent in 1960. In 1962, the buying-selling cooperatives were merged into the state enterprises under direct control of the Ministry of Internal Commerce.

By 1960 the process of Communist transformation of retail trade was completed when the socialist sector accounted for 91 percent of total domestic retail trade and 100 percent of foreign trade. Since 1965, however, because of the policy of wartime economic decentralization, the share of the socialist sector in domestic has been sharply reduced, and by 1971, accounted for only 32 percent of the total.

Side by side with the extension of government control in industry and trade, state freight transport grew at about the same speed as state wholesale trade. In 1957, up to half, and in 1963, 79 percent of the total volume of merchandise, was transported by the state enterprises. In passenger transport, expansion of control was somewhat slower in comparison with freight transport. Nevertheless, by 1963 passenger traffic carried by the state enterprises accounted for 75 percent of total traffic volume. By 1971, the total share of the socialist sector amounted to 83 percent as compared to 86 percent in 1960.

Finally, in order to regulate the 21 to 25 percent volume of traffic still carried by the private sector, the government adopted a principle of "three standardizations" (ba thong) of transportation; standardization of tariff, of traffic movement, and of distribution of merchandise.[28]

CONCLUSION

By 1960 Hanoi's leaders had effectively carried out a program of institutional changes modeled, step by step, after the Chinese and Soviet experience, to transform the North Vietnamese economy

TABLE 4.5

Share of State Enterprises and Cooperatives
in the Economy, by Sector, 1955-71
(in percentage)

Sector	1955	1960	1965	1968	1971
Industry	11	78	93	94	94
Agriculture	--	55	90	94	96
Construction	--	75	91	88	78
Transport	27	86	76	72	83
Domestic trade	21	93	82	79	57
Foreign trade	97	100	100	100	100

Source: DRVN, La Democratique Republique du Vietnam, (Hanoi: Editions En Langues Etrangeres, 1975), p. 64.

from mainly private ownership and a relatively free market economy into a collective-ownership-and-command economy.

In the rural sector, transformation was implemented through two broad programs: land reform and land collectivization. The first program, land reform, underwent two stages: the Land Rent Reduction Campaign (1949-53) and the Land Reform Campaign proper (1954-56). The second program, land collectivization, was carried out in three steps: first, the formation of work-exchange teams, including seasonal and year-round teams (1956-58); second, the establishment of low-rank agricultural cooperatives (1958-60); and third, with nearly 86 percent of rural families in cooperatives by 1960, and collectivization therefore considered complete, the advancement of cooperatives from low to high rank and the merging of small cooperatives into large ones.

According to the Chinese formula, high-rank cooperatives were to precede the establishment of the people's communes, in which the peasants live and work together in agromilitary camps. Until 1973, however, North Vietnam had stopped short of this final phase of collectivization, and so far the people's communes had not been introduced.[29]

As Hanoi's planners relied mainly on rural reform to increase agricultural production, failure of reform during 1954-56, as well as the weakening of economic incentives in the following period when collectivization was instituted, were among the main factors responsible for the depressed state of the rural economy during the five years after partition.

In the modern sector, the five-year period after 1954 was also the time when the process of consolidating state power was carried out with great speed. Side by side with agrarian reform, state enterprises were established in industry, mining, manufacturing, wholesale and retail trade, and transport.

In industry and mining, the Communist transformation at first was pursued as state capitalism in its lower form. In this phase, the government still permitted a certain degree of private enterprise to exist. Private enterprise was, however, subjected to the principle of "utilization, restoration, and control"; that is, the state only utilized it, but under strict control. Nevertheless, overridden by the fear of not having fulfilled their duty to advance international communism as quickly as possible, Hanoi's leaders in 1957 moved toward severe restrictions and complete elimination of private ownership in industry.

In manufacturing and handicrafts, the process of Communist transformation was carried out through expansion of cooperatives, as they were considered the best way to create collective ownership of the means of production.

In trade and transport, transformation was relatively slower. By the end of 1957 the private sector still shared about half of the volume of trade and transport. However, the trend was clearly set in that year toward complete Communist transformation. That process was completed in 1963 when private ownership was nearly eliminated.

Once the foundation for structural change had been firmly established, the government proceeded toward planning to build a truly Communist economy. The following chapter analyzes North Vietnam's two major development plans.

APPENDIX: SELECTIVE FEATURES OF COOPERATIVE MANAGEMENT*

1. Work-days (w. d.) required for different categories of labourers per year:

---

*Excerpted from Cuong Pham, "Management in Agricultural Cooperatives," in Etudes Vietnamiennes (English edition), no. 2 (Hanoi: Editions En Langues Etrangeres, 1964) pp. 108-26.

strong labourers	:240 w. d. for the men
average labourers	:230 w. d. for the women
mothers of many children	:150 w. d.
weak labourers	:100 w. d.

2. Organization of production brigades.

There are two forms of management in the co-operative: the management of the co-operative itself and that of the fixed production brigade. The fixed production brigade does not own any means of production, which are supplied by the co-operative, and its duty is to carry out the assigned tasks. It has no funds of its own. Each brigade elects a leader and two deputy leaders among whom there must be youths.

The formation of brigades is based on the requirements of production: kinds of crops, tilling methods, time allowed for completion of work, labour force necessary to fulfil the assigned task, geographical conditions and cadres' abilities for leadership.

In general, brigade membership varies between 30 and 40 (except for mountain regions). In agricultural co-operatives which also run other trades or grow such industrial crops as lacquer trees and tea plants this number may be higher but must not be more than 60.

3. Labour force employment.

To fulfil the production plan, the brigade must work out a plan of employment of labour force. It must clearly determine: The number of work-days allotted to each branch of production and each kind of work. The number of work-days devoted to civic work (large-scale water conservancy work in particular). The number of work-days devoted to the common tasks of the co-operative. The number of work points given to each family of co-op members as well as the way of determining the number of points given to families lacking in labour force.

4. Work point rating system.

Work point rating consists in classifying different kinds of work and determining the number of work days and work points for each category.

If the work requires great physical efforts, technical abilities and must be carried out in a short time, the remuneration will be relatively high. If not, it will be lower.

Once the classification is made the cooperative must determine the number of work points allotted to each category grade, giving 10 points to the average norm of work, that is, the work done by a person in a day with an average labour outlay and technique. The other grades are fixed on the basis of the average norm (below is an example, a table on classification and estimation of work at the Dai Phong co-operative).

Grade	Kind of Work	Work Points
1	Sweeping yards, cleaning duck-eggs, feeding pigs	6
2	Hunting the birds which destroy the crops, draining water, making ropes and baskets	7
3	Dressing and weeding secondary food crops catching fish, taking care of an incubator	8
4	Digging up the soil, scooping water, tending oxen	9
5	Ploughing, harrowing, transplanting, manuring, working a noria, reclaiming land, gather fire-wood, tending ducklings	10
6	Breeding ducks for eggs, felling timber	12
7	Worker specialised in hatching ducks	14
8	Sawing wood for making boats	15

5. Application of the rice-work contract

After a norm has been defined and labour graded into various categories, the co-op can apply the system of the 3-piece work contract:

Contract on production norms.

To determine production norms properly the co-op must first divide land according to soil into separate portions, then fix the norm for each portion and for the kind of plants grown taking into consideration the cultivation methods practiced. The norms fixed for each portion are assigned to the brigade.

Contract on the number of work-days and points to be assigned each brigade. The co-op bases itself on the quantity and quality of the land as well as on the kind of work entrusted to each brigade to determine the number of work-days and points. In the process of production, when working conditions change, the co-op must consequently alter the norm, and criteria concerning labour and the number of work-days. Thus, the number of work-days assigned to each brigade for every hectare of land may vary according to the circumstances. But in order to ensure to the co-op members belonging to various brigades a nearly equal income, the co-op must see to it that the number of work-days assigned to each brigade be nearly equal.

Contract on production costs. This contract ensures a good application of cultivation methods advocated by the co-op, avoids waste of seeds and funds as well as other unnecessary costs. It has also a compulsory character and makes it incumbent upon the brigades to achieve their tasks.

NOTES

1. D. T. Truyen and D. T. Vinh, Building an Independent National Economy in Vietnam (Hanoi: Foreign Languages Publishing House, 1964), p. 51.

2. See Pham Van Dong, Realisations et Taches Intermediates (Hanoi: Editions En Langues Etrangeres, 1955), pp. 64-65.

3. On the property rights of the land tiller, see Truong Chinh, Tang Cuong Doan Ket De Xay Dung Mot Nuoc Viet Nam Hoa Binh, Thong Nhat, Doc Lap, Dan Chu va Giau Manh (Strengthening unity to build a peaceful, unified, independent, democratic and prosperous Vietnam) (Hanoi: Su That Publisher, 1958), especially pp. 47-49.

4. The 25 percent rent reduction was promulgated by a decree on November 20, 1945. According to another decree, published on October 26, 1945, the reduction was 20 percent. See DRVN, Economic Institute, Kinh Te Vietnam (The economy of Vietnam) (Hanoi: Su That Publisher, 1960), p. 8; see also Nguyen Van Dam, Le Viet Nam En Marche (Hanoi: Edition En Langues Etrangeres, 1955), p. 97.

5. For more details, see Hoang Van Chi, From Colonialism to Communism (New York: Praeger, 1964), pp. 163-68.

6. The prime objective of the campaign, according to Van Chi was "the liquidation of the land-owning class and the subsequent establishment of the Proletarian dictatorship in the countryside" (ibid., p. 163).

7. Definition of the five categories of peasants is as follows:

"—a landlord is the one who does not work on his land, but lives on land rent and usury;

—a rich peasant tills his land himself and disposes of a surplus on which he exploits the farmhands' labour;

—a middle peasant lives solely by tilling his land without exploiting others;

—a land-poor peasant owns too little land to meet his needs and has to rent land from others or hire himself as farmhand;

—a landless peasant is most often a farmhand" (Truyen and Vinh, op. cit., p. 53.)

8. According to Truyen and Vinh (ibid., p. 52), the per-capita holding of land before and after land reform was as follows:

Category	Before Reform (square meters)	After Reform (square meters)
Landlord	6,499	1,010
Rich peasant	2,141	2,185
Middle peasant	1,151	1,665
Land-poor peasant	455	1,431
Farmhand	199	1,528

Results of the reform show that "land passed from landlords mainly to small and landless peasants, while the holdings of rich and middle-class peasants did not change substantially" (Theodore Shabad, "Economic Development in North Vietnam," Pacific Affairs, March 1958, p. 46).

9. On the various issues related to the "bloodbath" argument, see U.S. Congress, Senate, Committee on the Judiciary, The Human Cost of Communism in Vietnam—II; The Myth of No Bloodbath. Hearing before the Subcommittee to Investigate the Administration of the Internal Security Act and Other Internal Security Laws, 93d Cong., 1st sess., January 5, 1973 (Washington, D.C.: U.S. Government Printing Office, 1973).

10. For a detailed classification of different steps of collectivization, see Van Chi, op. cit., pp. 243-44.

11. For a general description of the cooperative and its evolution, see Louis P. F. Smith, The Evolution of Agricultural Cooperation (Oxford: Oxford University Press, 1961).

12. Except for the 5 percent of per-capita cultivable land that the peasant was allowed to retain under an incentive policy instituted in April 1959. See DRVN, Etudes Vietnemiennes, no. 13 (English edition) (Hanoi: Editions En Langues Etrangeres, 1967), p. 122.

13. See Hoang Van Chi, "Collectivization and Rice Production," in North Vietnam Today, ed. P. J. Honey (New York: Praeger, 1962), pp. 117-27.

14. Nghe An is also considered the very cradle of Vietnamese communism (Hoang Van Chi, From Colonialism to Communism, op. cit., pp. 224-25).

15. DRVN, Etudes Vietnamiennes, no. 13, op. cit., p. 10.

16. DRVN, So Lieu Thong Ke, (Statistical data), 1960-63 (Hanoi, 1963), p. 64.

17. For an elaboration on this point, see DRVN, Kinh Te Vietnam, op. cit., pp. 200-05.

18. Advantages of the peaceful transformation of the industrial sector were discussed by Tran Phuong, in Kinh Te Hoc Pho Thong (Popular economics), vol. 3 (Hanoi: Khoa Hoc Publisher, 1962), pp. 80-87.

19. Ibid., p. 89. See also Vo Nhan Tri, Croissance Economique de la Republic Democratique du Vietnam (1945-65) (Hanoi: Editions en Langues Etrangeres, 1967), pp. 221-23; Le Chau, Le VietNam Socialiste (Paris: Francois Maspero, 1966), pp. 254-56.

20. Truyen and Vinh, op. cit., pp. 75-76.

21. Ibid.

22. Stanley H. Cohn, Economic Development in the Soviet Union (Lexington, Mass.: D. C. Heath, 1970), p. 11.

23. G. Warren Nutter, The Growth of Industrial Production in the Soviet Union (Washington, D. C.: National Bureau of Economic Research, 1962), p. 185.

24. DRVN, Kinh Te Vietnam, op. cit., p. 207.

25. On the link between the capitalist class and the "hundred flowers" (nhan van giai pham), see ibid., pp. 209-11.

26. Ibid., p. 208.

27. For more details on trading arrangements, see Chau, op. cit., pp. 256-57.

28. Phuong, op. cit., p. 95.

29. Van Chi speculated on this point: "It would seem that this cautious attitude (of refraining from introducing communes into North Vietnam) has been advised by Khrushchev, who does not approve of the system of people's communes" (From Colonialism to Communism, op. cit., p. 242).

PART

III

THE PATH TOWARD UNBALANCED GROWTH

CHAPTER

5

ECONOMIC DEVELOPMENT PLANNING

As early as 1955 actual planning had already taken place with the establishment in North Vietnam of a National Planning Board and a Central Statistical Office. During the reconstruction program (1955-58), studies were undertaken to prepare a series of development plans to be executed once the reconstruction plan was completed.

North Vietnam started out with a three-year plan (1958-60), switched to a five-year plan (1961-65), and then mapped another five-year plan (1966-70) to bring about structural changes in the economy. The plans for the 1960s were described at the Seventh Session of the Central Committee of the Communist party:

> In the first period which lasts about ten years (beginning 1961) we must strive to bring about a basic change in our national economy, broadly apply mechanization and semi-mechanization in all branches, do away with the backward indigenous method and the present serious imbalance in [that] economy, ensure a balanced, harmonious and vigorous development of industry, agriculture and other economic branches with a view fundamentally to satisfying the everyday requirements of the people in food, clothing, housing, education and health protection; at the same time we will make preparations in all fields—materials, technique, cadres, workers, etc., to continue developing [the] economy all-sidedly, vigorously and steadily.[1]

As the air war began in the North in 1965, however, the second five-year plan was abandoned and was replaced with annual planning. This chapter analyzes the first two plans, covering the period 1958-65, which was the only period of formal and systematic planning in North Vietnam.

DEVELOPMENT STRATEGY

In faithful pursuit of the Marxist-Leninist ideology, the founders of the new regime applied the Chinese model of development, which had been based on the Russian plan, this strategy being accepted as the best possible approach for growth in North Vietnam.

As early as 1955 the blueprint for modernization had been worked out and was transmitted from the government to the National Assembly:

> With our extremely rich natural resources in the North, with the immense forces of our people and working class, with the disinterested aid of the brother countries, we have the necessary conditions to fulfill our plan of economic restoration and industrial development, and from this basis, to impel our industry, with a view to laying the foundations for the industrialization of our country . . .[2]

It seems, therefore, that there had never been a serious debate among North Vietnam's planners on the question of alternative approaches to development. Differences of opinion among them about priorities for investment were rather mild, occurring perhaps at lower levels of the policy-making structure. The director of planning, Le Thanh Nghi, summarized the situation this way:

> Many comrades have not yet fully realized that socialist industrialization is the central task of the entire Party and people throughout the period of transition to socialism; they have not duly realized the situation of the backward economy of our country and have not correctly assessed the achievements we have scored, therefore they become impatient, even pessimistic. Their understanding of the contents of socialist industrialization is still biased, departmentalist, they only attach importance to industry and belittle agriculture, or only think of developing agriculture without paying due attention to the development of industry and other economic branches. Owing to the

> fact that they take the priority development of heavy industry apart from the development of agriculture and light industry, a number of comrades have not realized that the task of heavy industry is to serve the development of agriculture, light industry and other economic branches. Vice-versa, there are comrades who only want industry to serve agriculture and fail to realize that agriculture must also serve industry.[3]

To the Northern planners, application of the Chinese-Soviet model of industrialization was a rational approach as their choice was influenced by the following factors:

1. The Chinese planning experts who came to North Vietnam as early as 1955 under a program of technical assistance were naturally eager to apply their own scheme of development.
2. The Chinese-Russian model was readily available and ideologically desirable.
3. The model had been tested in China, Russia, and in all other Eastern European countries with perhaps the exception of Yugoslavia.
4. Adoption of the Chinese-Russian model was desirable from the viewpoint of expected aid from both countries.

On the basis of the plan objectives as well as the pattern of planned investment, North Vietnam's two plans resemble strikingly the Russian and Chinese first five-year plans,[4] with the only difference being that the degree of decentralization was perhaps higher in North Vietnam since the provincial authorities were delegated the role that had been entrusted to the sovnarkhozes in Russia.[5] Like the Soviet Union, North Vietnam also started out with a three-year plan as a transitional step between reconstruction and formal planning. China on the other hand had launched its five-year plan immediately after its reconstruction program (1949-52). North Vietnamese planners appeared to have thought that a short period was needed to lay down a foundation for industrialization and for preparation of the planning process. Thus, North Vietnam's three-year plan may have been considered as an intermediate plan.

Implementation of the Chinese and Vietnamese first five-year plans proceeded in such a similar way that even the shifting of policies took place at about the same phase of each plan. China changed its strategy from emphasis on heavy industry to stressing the importance of agriculture and light industry during the last two years of its first five-year plan, in 1956 and 1957,[6] while North

Vietnam did the same in 1964 and 1965, the last two years of its first five-year plan.[7]

Failure to carry out the plans as originally envisaged pointed out the very difference between the initial economic conditions of Russia on the one hand and those of China and North Vietnam on the other. China's economy and, to a much greater extent, North Vietnam's economy were much more backward at the dawn of their respective plans in comparison with the economy of the Soviet Union.[8]

The vast majority of the population of North Vietnam were farmers living at subsistence level with very little left over even under strict austerity conditions imposed by the government in order to increase savings for capital formation and to support the new army of industrial workers. Thus, while Stalin was able to extract a continuous stream of savings from the Russian agricultural sector to finance his industrialization plan, the Vietnamese Communists were unable to do so in the face of a most serious constraint—an inadequate supply of food.

PLAN OBJECTIVES

In general terms both the three-year plan and the first five-year plan, although stated in different ways, implied similar objectives. North Vietnam's main strategy of development as adopted from the Chinese model was to achieve rapid increase of output through unbalanced growth favoring heavy industry. Heavy industry was defined as including electricity and fuel, engineering, metallurgy, chemicals, mining, construction materials, and timber; the most important branches were engineering, machinery, and electricity. Light industry included food processing, textiles, wood processing, building materials, transportation, communications, and other consumer goods.

In addition to the economic reason for favoring industrialization, Hanoi's planners also viewed industrialization as the most effective way to bring about complete Communist transformation of North Vietnam's economy and to strengthen the defense sector.[9] From the ideological standpoint, industrialization encourages growth in numbers of factory workers, thus making for the emergence of a Marxian proletariat from a mainly agricultural society.

Concerning the objectives of the plans, although the three-year plan stated only three as compared with five in the five-year plan, the two plans embodied generally the same objectives. They can be summarized as follows.

First, in principle, the three-year plan emphasized agriculture and manufacturing, while the five-year plan gave priority to heavy industry.[10] However, in terms of planned investment and growth rates, agriculture did not actually receive priority, even in the three-year plan. The policy of unbalanced growth favoring industry was stated formally in a resolution of the Third National Party Congress held in September 1960 on the eve of the five-year plan.[11] In that plan a growth rate was envisaged for industry twice as high as that for agriculture, and investment allocated to industry was three times greater.

The second objective of the plans was to accelerate the socialist transformation of the economy.[12] In practice, this objective was carried out through the application of collectivization in agriculture and through the institution of the "higher form of state capitalism" in industry. Compared with the five-year plan, the three-year plan put more emphasis on agricultural collectivization, very likely because of the planners' conviction that collectivization was not only necessary from the viewpoint of ideology, but also because it was seen as the best way to raise agricultural productivity as well as to increase the contribution of agriculture to industry. The growth of this contribution is required in order to support the growing number of workers required for industrialization.

The third objective was to raise the cultural (actually the ideological) level and well-being of the masses. Stress was laid on the training of skilled workers and managers, and on promotion of scientific research. The five-year plan subdivided the third objective into two, dealing separately with the cultural level and the living standard.[13]

The final theoretical objective of the plans, but very likely the most important ultimate objective, was the consolidation of national defense and security. In the three-year plan, national defense was given minor treatment and was included in the third objective, while in the five-year plan it was covered separately.[14]

Plan Growth Targets

Both plans established ambitious targets for growth rates in all sectors. However, it appears that the experience obtained from the implementation of the three-year plan had resulted in some moderation in planning for the five-year plan. Table 5.1, which was constructed on the basis of data from various official and unofficial sources, shows that all the planned targets of the five-year plan were lower than those for the previous plan.

The five-year plan was considered the most important one in that it was the beginning of systematic planning for complete mechanization of the economy and was also the first one to give priority to heavy industry. Yet the planned annual rate of growth of industry and

TABLE 5.1

Development Plans: Planned and Realized Growth Rates, by Sector
(percentage)

Sector	Planned, 1958-60		Realized, 1958-60		Planned, 1961-65		Realized, 1961-65
	Total Plan Period	Average Per Annum	Total Plan Period	Average Per Annum	Total Plan Period	Average Per Annum	Average Per Annum
Industry and handicrafts	86.4	23.1	81.2	21.9	148.0	20.0[c]	14.2[e]
Modern industry	139.0	33.9	157.5	37.1	—	—	—
Group A[a]	159.0	37.2	168.0	39.0	215.0	25.8	21.1[f]
Group B[b]	69.1	19.2	60.4	15.8	110.0	16.0	12.2[f]
State industry	197.0	42.8	205.0	45.0	199.0	24.5	22.3[f]
Handicrafts	34.0	10.2	29.2	8.9	—	—	5.8[f]
Agriculture	41.8	12.2	13.5	4.4	61.0	10.0[d]	4.1[e]
Commerce (retail sales)	40.0	11.9	17.8	5.6	—	—	6.1[f]
Transport (merchandise, million tons per kilometer)	138.0	33.5	140.0	34.0	—	—	11.4[f]
Other targets							
Productivity of labor	59.0	16.8	48.0	19.2	30.0	5.4	—
Real income of workers in industry (and agriculture)	25.0	7.2	—	—	—	—	—
Education (number of students)							
General	80.0	26.7	48.0	16.0	100.0	14.9	—
Higher	100.0	33.3	63.0	21.0	200.0	24.5	—
Technical training	230.0	76.7	145.0	48.3	150.0	20.2	—

[a]Also referred to as capital-goods industry or heavy industry (including public utilities, mining, engineering, constriction materials, rubber, and timber products).

[b]Also referred to as consumer-goods industry or light industry (including textiles and textile products, food processing, leather goods, "cultural articles," and other consumer products).

[c]It was later revised to 17 percent.

[d]It was later revised to 6.5 percent.

[e]For 1961-64.

[f]For 1961-63.

Sources: Reconstructed North Vietnamese data, and estimated, based on DRVN, Nghien Cuu Kinh Te, no. 1 (Hanoi, 1961), pp. 3, 60, and no. 2, (1961), pp. 51-54; Hoc Tap, no. 5, 1961, pp. 23-31; Le Chau, Le Viet-Nam Socialiste, Une Economie de Transition (Paris: Francois Maspero, 1966), pp. 312, 316, 319 (hereafter cited as Le Viet-Nam Socialiste); Vo Nhan Tri, Croissance Economique de la Republique Democratique du Vietnam (1945-65) (Hanoi: Editions En Langues Etrangeres, 1967), pp. 290-92, 420, 435, 540; Leon Lavallee, L'Economie du Nord Vietnam (Paris: Centre d'Etude et de Recherche Marxiste, 1971), p. 24.

handicrafts was 20 percent and that for heavy industry (group A), 25.8 percent as compared with 23.1 and 37.2 percent, respectively, in the three-year plan. Light industry (group B) was projected to grow at a much smaller rate in both plans, 19.2 percent in the three-year plan and 16 percent in the five-year plan. The former also established a separate target for handicrafts (8.9 percent annually).

In all, the value of industrial and handicraft production was envisaged to account for 41.7 percent and 48.5 percent of total national output in 1960 and 1965, respectively, implying a decrease in agriculture's share of GNP by the final years of each plan.

Both plans stressed the important role of agriculture in economic development as well as the ability "to transform and develop agriculture simultaneously with industry." The annual growth rate in agriculture was projected at 12.2 percent for the three-year Plan and 10.0 percent for the five-year plan. The latter was revised to 6.5 percent during 1963. Agriculture-growth rates were thus about half of those envisaged for industry in both plans. Within the agricultural sector, the three-year plan emphasized the overriding need to increase rice output in order to solve the food-shortage problem, while the five-year plan also stressed the need to push the production of nonstaple crops such as peanuts, sugar, tea, coffee, cotton, rubber, and cigarettes to supply industry, and for export.

In terms of quantity, the three-year plan called for rice production to increase from 3.95 million tons in 1957 to 7.6 million tons in 1960.[15] The five-year plan's rice target of 7 million tons by 1965 was less ambitious than that of the three-year plan, and the target was further revised to 5 million tons.[16] Cultivation of non-staple foods was expected to represent 10 percent of total area under cultivation by 1965.[17]

In industry, the three-year plan envisaged a target of 271 million kilowatt hours for electric power by 1960, against 123 million kilowatt hours in 1957, and a target of 2.7 million tons of coal, against 1.1 million in 1957. Cement production was expected to reach 450,000 tons in 1960, compared with 165,000 tons in 1957, and textile output was projected to be 82 million meters, against 39 million meters (see Table 5.2).

The five-year plan included the following main industrial targets:[18]

1. Building of a diesel-engine workshop, a shipyard, and a workshop to produce metal-cutting machines, spare parts for motor-cars.

TABLE 5.2

Development Plans: Selected Output Targets and Realization

Product	1957	1960 Target	1960 Realized	1960	1965 Target	1965 Realized
Paddy (million tons)	3.9	7.6	4.2[a]	4.2	7.0[b]	4.5[b]
Paddy per capita (kilograms)	271	500	227	227	—	243
Electricity (million kilowatt hours)	123	271	254	254	660	678
Coal (million tons)	1.1	2.7	2.4	2.4	5.0	4.0
Cement (thousand tons)	165	450	410	410	710	640
Textiles (million meters)	39	82	86	86	125	134

[a]Official figure.

[b]The 7-million-ton target was later revised to include all foodstuffs, of which only 70 percent (5 million tons) was rice.

Sources: DRVN, Plan Triennal de Developpement et de Transformation de l'Economie et de Developpement de la Culture (Hanoi: Editions En Langues Etrangeres, 1959), pp. 48, 55, 67; Le Chau, Le Viet-Nam Socialiste (Paris: Francois Maspero, 1966), pp. 320, 322; Vo Nhan Tri, Croissance Economique de la Republique Democratique du Vietnam (1945-65) (Hanoi: Editions En Langues Etrangeres, 1967), pp. 293, 322; D. T. Truyen and P. T. Vinh, Building an Independent National Economy in Vietnam (Hanoi: Foreign Languages Publishing House, 1964), p. 107; Leon Lavallee, L'Economie du Nord Vietnam (Paris: Centre d'Etude et de Recherche Marxiste, 1971), p. 50; Cuong Pham, "Five Tons Per Hectare?" in Etudes Vietnamiennes (English Edition) (Hanoi: Editions En Langues Etrangeres, 1965), p. 82.

2. Electricity: 660 million kilowatt hours in 1965, or 148 percent over 1960.

3. Completion of the Thai Nguyen steel and iron complex, to turn out 200,000 tons of steel in 1966. The first melting, scheduled for 1964, took place in November 1963.

4. Coal: 5 million tons in 1965, or double the 1960 figure.

5. Apatite: 14 million tons in 1965 as against 490,000 tons in 1960.

6. A superphosphate plant having a yearly capacity of 100,000 tons was completed in 1963 and a nitrogen-fertilizer mill was to be completed by the end of 1965.

7. Cement: 710,000 tons in 1965 as against 408,000 tons in 1960.

8. Textiles: 125 million meters of cotton fabric in 1965 as against 76 million in 1960; 8 million meters of cloth of other fibers, 1.5 million meters of silk; 32 million meters of mosquito netting.

9. Wood: 1.31 million cubic meters in 1965 as against 753,000 in 1960.

Among other targets, productivity of labor was expected to rise at an annual rate of 16.8 percent during the three-year plan as compared with 5.4 percent during the five-year plan. This was in spite of the fact that during the latter, when industrialization was expected to progress steadily, labor productivity would have to be higher than in the previous period.

The setting up of growth targets for the five-year plan at a level significantly lower than those for the three-year plan in all sectors may be explained not only in terms of experience from the three-year plan, but also in terms of the influence of Chinese planning at the time: North Vietnam's three-year plan was implemented precisely at the same time as the Great Leap Forward in China (1958-60) and its first five-year plan coincided with the period of development in China under the conditions of "caution, moderation and policy adjustment" (1961-65).[19] No doubt, the pessimistic spirit after the Great Leap Forward in China as well as the cautious attitude of Chinese planners during 1961-65 had certainly influenced the North Vietnamese planners in establishing growth targets for their five-year plan. In China, Premier Chou En-Lai defined a 20 percent increase in industrial output as a "leap forward," a 25 percent increase as a "great leap forward," and a 30 percent increase as an "exceptionally great leap forward."[20] On those criteria, North Vietnam's targets for increase in industrial output during the three-year plan may be characterized as a "great leap forward," while those in the five-year plan were merely a "leap forward."

Level and Structure of Investment

Total public-investment outlays were envisaged at 1.48 billion dong (U.S. $420 million) for the three-year plan, and 5.1 billion dong (U.S. $1,445 million) for the five-year plan (see Table 5.3). Compared with the level of investment realized during the period of reconstruction, the three-year plan's annual average of 495 million dong (U.S. $140 million) represented an increase of over 38 percent.

On the basis of the estimated national income, the planned level of public investment in the three-year plan implied an average ratio of gross investment to gross domestic product (GDP) of about 14 percent per annum, or about the standard ratio required for reaching economic "take-off."[21] Public investment planned for the five-year plan, averaging 1.02 billion dong per annum (U.S. $289 million), was more than double the planned average for the previous plan, implying a rate of gross (public) investment of about 22 percent of GDP (the rate is calculated in terms of the average GNP for 1961-65, as shown in Chapter 8.).

In addition to investment financed by the central-government budget, there were investments made by regional authorities and cooperatives that were not included in the amount of investment in the plans. It was estimated that cooperatives would invest 5 percent of their revenues per year during the three-year plan and between 12 and 15 percent during the five-year plan.[22] Also important was the part of investment planned in the form of labor investment. Under this scheme, the rural population was required to participate in the plan by contributing a number of days per year to work on local projects, especially moving earth to build irrigation networks, roads, and public buildings. The three-year plan called for a total labor investment of 127 million workdays per year.[23] The five-year plan required each peasant to devote 70 days per year to "secondary undertakings," that is, work on nonagricultural projects.[24] Taking into account local investment and investment by labor, the planned ratio of investment to GDP may have been somewhere around 18 percent for the three-year plan and 27 percent for the five-year plan (assuming that local investment and investment by labor amounted to 5 percent of GDP). The estimated investment ratio in the three-year plan was therefore about the same as that in the Chinese first five-year plan, while the ratio envisaged for the North Vietnamese five-year plan was on the order of that of the Russian first five-year plan.[25] Considering the very low level of income of the masses in the rural sector in North Vietnam, however, the planned level of investment was exceptionally high.

The sectoral distribution of investment in the two plans as shown in Table 5.3 suggests that Hanoi's planners were determined

TABLE 5.3

Development Plans: Sectoral Distribution of Planned Public Investment
(in millions of dong)

Sector	Original Plan, 1958-60			Revised Plan, 1958-60			Five-Year Plan, 1961-65		
	Percent of Total	Amount	Annual Average	Percent of Total	Amount[a]	Annual Average	Percent of Total[b]	Amount[c]	Annual Average
Productive	82.9	1,230	410	81.9	1,215	405	88.7	4,524	905
Industry	46.5	690	230	41.1	610	203	48.5	2,473	495
Group A[e]	—	—	—	27.4	406	135	38.3	1,954	391
Group B[e]	—	—	—	13.7	204	68	10.2	519	104
Agriculture	10.5	260	87	12.6	187	62	21.8	1,112	222
Transport and communications	17.5	156	52	20.6	306	102	18.4 (Transport and communications, Commerce and construction combined)	938	188
Commerce and construction	8.4	125	42	7.6	113	38			
Nonproductive sector	17.1	254	85	18.1	269	90	11.3	576	115
Education, culture, public health	5.8	86	29	8.6 (Education, Research, Housing combined)	128	43	4.3[d]	219	44
Research	2.2	33	11						
Housing	4.7	69	23	—	—	—	—	—	—
Administrative services	2.5	37	12	9.5	141	47	7.0	357	71
Utilities	1.9	28	9	—	—	—	—	—	—
Grand Total	100.0	1,484	495	100.0	1,484	495	100.0	5,100	1,020
U.S. equivalent (in millions of dollars)		420	140		420	140		1,445	289

[a]Based on the assumption that the total planned level of investment, 1,484 million dong remained unchanged, that is, only percentage distribution was revised.
[b]Equals average of distribution of new investment fixed for 1962-64.
[c]On the basis of total investment for the whole plan being 5,100 million dong.
[d]Average of 1962 and 1963 only.
[e]For the three-year plan, group A received 67 percent of investment allocated for total industry, and group B, 33 percent. For the five-year plan, group A received 79 percent and group B, 21 percent.

Note: Individual items may not add up to total because of rounding.

Sources: DRVN, Plan Triennal de Developpement et de Transformation de l'Economie et de Developpement de la Culture (Hanoi: Editions En Langues Etrangeres, 1959), pp. 76-77; So Lieu Thong Ke, 1955-59 (Hanoi, 1960), pp. 97-101; 1960-63, p. 73; Le Chau, Le Viet-Nam Socialiste (Paris: Francois Maspero, 1966), pp. 316; Vo Nhan Tri, Croissance Economique de la Republique Democratique du Vietnam (1945-65) (Hanoi: Editions En Langues Etrangeres, 1967), pp. 543, 555; Nghien Cuu Kinh Te, no. 48 (1969), p. 13.

to industrialize the economy at a speed as fast as that in China.[26] For investment-priority purposes, the economy was divided into two sectors: the productive sector and the nonproductive sector. The productive sector included, in decreasing order of importance, industry, agriculture, transport and communications, commerce, and building. The nonproductive sector included education, public health, research, housing, administrative services, and public utilities. In both plans the productive sector accounted for most of investment, leaving only 17 percent for the nonproductive sector in the three-year plan and a little over 11 percent in the five-year plan.

Nearly half of the total amount in both plans was to be invested in industry, with some downward adjustment made during the three-year plan from 47 percent to 41 percent. Within the industrial sector, group A accounted for about two-thirds of investment. Within group A, electricity and mining received the highest shares of investment, as did textiles and food processing within group B.

The original allocation of investment for agriculture in the three-year plan was only 10. 5 percent, compared with 46. 5 percent for industry. The planners apparently counted heavily on labor investment by the rural population and on collectivization to increase agriculture output. However, as the rural sector continued to deteriorate and the food situation became more and more acute, investment allocated for agriculture was revised upward to 12. 6 percent after the three-year plan was launched.[27] In the five-year plan, agriculture's share of total investment was 21. 8 percent, or more than double the original level in the three-year plan. Within agriculture, over half of investment was reserved for irrigation and water-control projects,[28] a fact which reflects the planners' overriding concern for the vulnerability of North Vietnam's agriculture (as discussed in Chapter 2).

Investment in transport, communications, commerce, and construction combined, originally envisaged at 26 percent in the three-year plan, was later increased to 28. 2 percent as compared with 18. 4 percent in the five-year plan. Most of this investment was to strengthen the enterprise of the state sector, especially to increase merchandise transport and retail trade carried out by the state enterprises. The growth of state enterprises was analyzed in Chapter 4.

Financing Investment

In light of the very low level of income and thus of domestic savings in North Vietnam, the planners seemed to rely heavily on foreign resources in order to finance the development plans. "In our Three Year Plan, the construction of the major part of medium

TABLE 5.4

Financing of Planned Investment, 1958-65
(in millions of dong)

Source of Financing	Total Amount	Annual Average	Percent of Total
	Three-Year Plan, 1958-60		
Domestic	570[a]	190	38
External	914[b]	305	62
Grants	826[b]	275	56
Loans	88[b]	29	6
Total	1,484	495	100
U. S. equivalent (in millions of dollars)	420	140	
	Five-Year Plan, 1961-65		
Domestic	3,348	668	66
External	1,762	352	34
Grants	89	18	2
Loans	1,673	334	32
Total	5,100	1,020	100
U. S. equivalent (in millions of dollars)	1,445	289	

[a]Estimated as residual of total after allowing for external resources.

[b]Estimated.

Note: Individual items may not add to totals because of rounding.

Sources: Tables 5.1 and 5.3; Vo Nhan Tri, *Croissance Economique de la Republique Democratique du Vietnam (1945-65)* (Hanoi: Editions En Langues Etrangeres, 1967), p. 555; William Kaye, "A Bowl of Rice Divided," in *North Vietnam Today*, ed. P. J. Honey (New York: Praeger, 1962), p. 115.

and heavy industries is to be realized by assistance from fraternal countries," declared Nguyen duy Trinh, then president of the State

Planning Commission.[29] On the eve of the five-year plan, the party's Third National Congress held in 1960 declared: "As our economy is backward, agricultural and is advancing toward socialism, by-passing the state of capitalist development, the aid of the socialist camp is indispensable and is a very important factor."[30]

The two plans specified that the Soviet Union would help finance projects in electric power, mining, and fertilizer plants; that China would assist in the reconstruction of railroads, building of iron and steel mills, oil refineries, and in consumer goods manufacturing; and East European countries would help mainly in building activities.

The plans did not specify the sources of financing by country. However, on the basis of fragmentary information on foreign loans and grants, the sources of financing, by domestic and foreign origin, may be estimated; these estimates are shown in Table 5.4.

According to these estimates, 62 percent of the total financial requirements for the three-year plan was expected to come from external sources, of which over 89 percent (826 million dong) was in the form of grants and the rest, 11 percent (88 million dong), in long-term loans. The bulk of both grants and loans was provided by the Soviet Union, 469 million dong, and by China, 353 million dong. Between 1955 and 1960, the Soviet Union provided 400 million old rubles (U.S. $100 million) in grants and 510 million old rubles (U.S. $128 million) in loans, while China provided 900 million yuan (U.S. $375 million) in grants and 300 million yuan (U.S. $125 million) in loans.[31]

Foreign resources envisaged for the five-year plan were 1,762 million dong (U.S. $499 million), an average of 352 million dong per annum as compared with 305 million dong in the three-year plan, representing an increase of 15 percent. In terms of total financial requirements, however, foreign resources represented 34 percent, a substantial reduction in comparison with the share of foreign resources in the three-year plan. Most of the foreign resources provided for in the three-year plan were in the form of grants, while the entire amount supporting the five-year plan was in the form of loans. In 1960 alone the Soviet Union lent 430 million old rubles (U.S. $108 million) and China lent 142 million yuan (U.S. $59 million). Interest on the loans was exceptionally low, ranging from 1 to 2 percent per annum.[32]

Domestic resources, which amounted to 38 percent and 66 percent in the three-year and five-year plans, respectively, were to come entirely from the budget. According to official data, nearly 59 percent of budgetary expenditures during the three-year plan and 61 percent during the five-year plan were allocated for economic affairs.[33] In the budget, the government's most important budgetary revenue during the reconstruction period (1955-57) came from taxes, while in the three-year plan it came from "contributions from the state enterprises."*

State-enterprise contributions to the budget increased from 7 percent of total revenues in 1955 to 37 percent in 1958. During the three-year plan, average revenues coming from the state sector were 49 percent, and during the five-year plan, 56 percent. The fast growth of government revenues coming from the state enterprises is explained by the fact that by 1960 state enterprises had taken over most of the economy, as noted in the preceding chapter. The increasingly important role of state-enterprise revenue contributions to government receipts indicates that increasingly the high rate of capital formation was directly financed by them.

Looking at fiscal revenues by origin, about two-thirds of total government tax revenues during the plans were from the modern sector and one-third from the agricultural sector.

CONCLUSION: ACHIEVEMENTS UNDER THE PLANS

When North Vietnam entered the final year of its first five-year plan, large-scale fighting flared in the South, and American bombing over the North began. Hanoi's decision to carry out the war effort certainly prevented several ongoing projects from being completed and a number of new ones from being launched. Information is not available in the form of comprehensive evaluations or adequate annual progress reports on the plans to permit an accurate and meaningful appraisal of the plans' performance. Results of the plans as analyzed in this section are based mainly on fragmentary North Vietnamese sources, which are often ambiguous and inconsistent. As such, it is inevitable that they are rather inflated. The reader may keep in mind the "30 to 40 percent discount" rule in order to arrive at a more accurate picture. Furthermore, since agricultural and industrial production during 1955-65 will be dealt with separately in Chapters 7 and 8, respectively, this section only appraises some macro targets.

---

*These so-called contributions should not be considered as profits of the state enterprise, but rather a form of "forced saving" by the population going to the state's budget through the state enterprise.

Planned versus Actual Growth Rates

In attempting to adapt the Sino-Soviet model of development to North Vietnam's needs, it seems that Hanoi's planners were trapped into setting production targets rather too high for the predominantly agricultural and backward economy of the country. Thus, during the plans, targets—especially those set for agriculture—were often revised downward.

Since it is not possible to cover all the plans' targets together with their numerous revisions, the following discussion deals mainly with targets originally set either in the plans or by various officials as general policy guidelines.

Industry

As shown in Table 5.1, realized growth rates as claimed by North Vietnamese planners in the industrial and handicrafts sector just about met the targets set forth in the three-year plan, while in the five-year plan they averaged 71 percent of the annual targets.

The single factor most responsible for the allegedly satisfactory growth rate in this sector during the three-year plan was the fast rate of expansion of mining activities, electrical output, and other lines of production classified as industry group A, which increased at an annual average of 39 percent against 37 percent targeted. Light industry grew at a much slower rate, 16 percent against 19 percent targeted, but state industry surpassed the 43 percent planned, increasing at an annual average rate of 45 percent.

No official results on growth rates were given for the five-year plan. Fragmentary results for 1961-64 indicate that, compared with the results claimed for the three-year plan, actual growth rates realized for all branches in the modern sector were much lower than targeted. Growth for industry and handicrafts realized for 1961-64 averaged only 14 percent per annum against 20 percent targeted, in spite of "brilliant achievements" claimed for capital construction during 1964. The average annual growth rate obtained for industry (group A) was 21 percent during 1961-63 as compared with 26 percent originally planned, while that for light industry (group B) was 12 percent compared with 16 percent envisaged.

In terms of actual output, officially claimed results (as summarized in Table 5.2) show that electric energy reached 254 million kilowatt hours in 1960, the last year of the three-year plan, or more than double the 1957 level. As such, electrical output was only 6 percent less than the planned level. Coal production was also more than double the 1957 level, reaching 2.4 million tons in 1960 as compared with 2.7 million tons targeted. Cement output, 410,000 tons in 1960, was 9 percent less than planned. Textile production on the other hand surpassed its target by 5 percent, reaching 86 million meters in 1960.

In the five-year plan electrical output in the final year (1965) was reported to have reached 678 million kilowatts, or 3 percent higher than the planned level. However, both coal and cement production were significantly lower than their targets. In 1965, coal output was 4 million tons compared to 5 million tons planned, or 20 percent lower while cement output was 640,000 tons against 710,000 tons envisaged, or 10 percent less.

Agriculture

If North Vietnam's planners claimed satisfactory results for the modern sector, they also admitted disappointment in agriculture. In both plans, the original targets set for agriculture had to be revised downward.

In the three-year plan, the realized growth rate averaged only slightly more than 4 percent per annum rather than the ambitious target of over 12 percent. Results for the first four years of the five-year plan showed almost the same growth rate, 4 percent annually as against 10 percent targeted. The failure of agriculture to meet the planned growth rates was due mainly to poor performance in paddy production. The planners often blamed disappointing crop yields on "natural calamities." It is quite true that, in spite of tremendous amounts of labor invested in irrigation and water control, the Red River seemed to remain as uncontrollable in 1965 as ever before, and floods still devastated the delta frequently. Nevertheless, to blame poor agricultural performance entirely on nature is a matter subject to dispute. Certainly one of the factors behind agricultural failure was the planners' placing too much emphasis on industry and their reliance on land reform and collectivization as a driving force to increase agricultural productivity. As already noted, agriculture's share of total investment in the three-year plan was only 22 percent of that provided for industry, and its share in the five-year plan was less than half of that allocated for industry.

In terms of production target, paddy output in the final year of the three-year plan was 4.2 million tons as compared with 7.6 million tons planned, or 45 percent lower. In the final year of the five-year plan, paddy output was 4.5 million tons against the original target of 7 million tons, or 36 percent smaller.

Development during the plans in terms of quantitative trends is analyzed in the next chapter.

NOTES

1. Le Thanh Nghi, "Tasks and Trends of Socialist Industrialization in our Country," in DRVN, The Road to Happiness and Prosperity (Hanoi: Foreign Languages Publishing House, 1963), pp. 150-51.

2. DRVN, Report to the National Assembly, September 1955, in D. T. Truyen and P. T. Vinh, Building an Independent National Economy in Vietnam (Hanoi: Foreign Languages Publishing House, 1964), p. 78.

3. Le Thanh Nghi, "To Thoroughly Grasp the Party's Line on Socialist Industrialization," in DRVN, The Road to Happiness and Prosperity, op. cit., pp. 139-40.

4. The three plans (Soviet, Chinese, and North Vietnamese) all emphasized heavy industry (especially electric power, chemicals, and machinery). The planned percentage share of investment for industry in the Russian and Chinese first five-year plans and in North Vietnam's two plans was as follows: Russian first five-year plan (1928-32)—40.9 percent; Chinese first five-year plan (1952-57)—47.9 percent; North Vietnamese three-year plan (1958-60)—46.5 percent; North Vietnamese first five-year plan (1961-65)—48.5 percent. See K. C. Yeh, "Soviet and Communist Chinese Industrialization Strategies," in Soviet and Chinese Communism, ed. Donald W. Treadgold (Seattle: 1967), University of Washington Press, pp. 330-34; Abrams Bergson, The Economics of Soviet Planning, (New Haven: Yale University Press, 1964), p. 308.

5. Le Chau, Le Viet-Nam Socialiste (Paris: Francois Maspero, 1966), p. 309.

6. On the liberal interlude in China (October 1956-June 1957) and the change in policy, see Jan S. Prebyla, The Political Economy of Communist China (Scranton: International Textbook Co., 1970), pp. 224-26.

7. In 1963, amidst the agricultural crisis, Hanoi gave official directives to focus economic effort on agriculture and light industry rather than on heavy industry. See Far Eastern Economic Review, 1964 Yearbook (Hong Kong, 1964), p. 238.

8. For a comparison of the Chinese and Russian economies relevant to this point, see Yeh, op. cit., p. 347.

9. See Thanh Nghi, in DRVN, The Road to Happiness and Prosperity, op. cit., p. 156.

10. The most important objective of the three-year plan was "To intensify agricultural and industrial production taking agriculture as the main link [of development] and with an objective to solve the problem of foodstuffs. At the same time, to attach great importance to industrial production in order to increase to a maximum the means

of production and to solve—to a large extent—the problem of [shortage of] consumer goods" (DRVN, Plan Triennal de Developpement et de Transformation de l'Economie et de Developpement de la Culture, (Hanoi: Editions En Langues Etrangeres, 1959), pp. 35-36. [hereafter cited as Plan Triennal]). The first objective of the five-year plan was to "develop industry and agriculture, to realize in priority the development of heavy industry parallel to development of light industry and transport, and to the consolidation of state and cooperative commerce" (Chau, op. cit., p. 326).

11. The Third National Congress of the Workers' Party held in Hanoi on September 7, 1960, was considered the most important congress of the Vietnamese Communists. The congress was observed by delegations from eleven Communist countries and seven non-Communist countries (Chau, op. cit., p. 325). In its resolution, the Third Congress states: "When socialist transformation has achieved decisive successes, we will take socialist construction as the central task, carry out socialist industrialization, and at the same time complete socialist transformation" (DRVN, The Road to Happiness and Prosperity, op. cit., p. 13.)

12. DRVN, Plan Triennal, op. cit., p. 29; Chau, op. cit., p. 326.

13. Ibid.

14. Ibid.

15. Plan Triennal, op. cit., p. 48.

16. Original target was 9.5 million tons of foodstuff, of which 7 million tons were paddy (Resolution of the Fifth Plenum of the Party Central Committee, July 1961, on the development of agriculture during the five-year plan, in DRVN, Offensive Centre Le Pauvrete et le Retard Economique, [Hanoi: Editions En Langues Etrangeres, 1963], p. 22). The target was later stated as 7 million tons of foodstuffs, of which 5 million tons (70 percent) were paddy. See Cuong Pham, "Five Tons Per Hectare?" in DRVN, Etudes Vietnamiennes (English Edition) (Hanoi: Editions En Langues Etrangeres, 1965), p. 82.

17. DRVN, Offensive Contre La Pauvrete et le Retard Economique, op. cit., p. 23.

18. Truyen and Vinh, op. cit., p. 107.

19. Opening the 1961-65 period's "Walk slow" policy, Liu Shao-ch'i declared in 1961: "Our economy is approaching the brink of collapse. I propose to call the present a period of emergency." The period of caution and moderation ended with an upheaval—the Cultural Revolution—that reversed the policy of moderation. See Prebyla, op. cit., pp. 341-46.

20. Yeh, op. cit., p. 350.

21. For a definition of "take-off," see W. W. Rostow, The Process of Economic Growth, 2d ed., (New York: Norton and Co., 1962), p. 284.

22. Chau, op. cit., p. 336.

23. Plan Triennal, op. cit., p. 79.

24. Chau, op. cit., p. 337.

25. For estimates on Chinese investment ratios, see William W. Hollister, "Trends in Capital Formation in Communist China," in An Economic Profile of Mainland China, vol. I, Studies prepared for the Joint Economic Committee of the United States, (Washington, D. C.: U. S. Government Printing Office, 1967), pp. 124-27. For estimates on Russia's investment ratios, see Bergson, op. cit., p. 308; K. C. Yeh, "Capital Formation," in Alexander Eckstein et al., Economic Trends in Communist China (Chicago: Aldine Publishing Co., 1968), especially pp. 510-11.

26. To compare with Chinese planning targets, see Prebyla, op. cit., pp. 339 and 370.

27. Le Chau, op. cit., p. 316.

28. DRVN, So Lieu Thong Ke, (Statistical data), 1955-59 (Hanoi, 1960), p. 101; Nghia Cuu Kinh Te (Economic research), no. 39, (1967): p. 13.

29. Plan Triennal, op. cit., p. 120.

30. Truyen and Vinh, op. cit., p. 88.

31. Vo Nhan Tri, Croissance Economique de la Republique Democratique du Vietnam (1945-65), (Hanoi: Editions En Langues Etrangeres, 1967), pp. 554-55. These amounts were provided to support the plans alone.

32. Ibid., p. 556.

33. See ibid., pp. 349-50, 353, 465.

CHAPTER

6

QUANTITATIVE TRENDS

CONCEPTS AND METHODOLOGY

With the exception of fragmentary information on national income for 1957-59 and on a number of growth indexes, North Vietnam has not published any systematic series on national accounts. In addition to the absence of official statistics, the conceptual discrepancies between Western and Communist measurements of these aggregates renders any attempt at deriving an accurate time series to analyze economic trends extremely difficult.

Assisted by Russian statisticians, North Vietnam adopted in the late 1950s a statistical system similar to that of the Soviets. The discrepancy in measurement, say, of national income, between the Communist and Western methodology is due to the fact that the Communist national accounting excludes the so-called nonproductive services from national products while, at times, it includes the intermediate products. As such, it often involved the problem of double accounting. For example, the value of a bicycle and its components, such as tires, may be all summed up. The double-accounting problem is perhaps more serious in industry than in agriculture. The nonproductive services include such components as administration, finance and credit, private housing, education, public health, entertainment, and personal services.[1]

To present for the first time a somewhat meaningful series of national accounts for North Vietnam, painstaking effort was taken to derive the figures for the gross material product (GMP) (which is different from the Western concept of gross national product) and national income as shown in Tables 6.1 to 6.4. These series are, therefore, really more indexes of relative performance over a period

TABLE 6.1

Estimated Gross Material Product, 1955-70
(at constant prices and in millions of dong)

	Component of GMP								
Year	Agriculture[a]	Industry and Handicrafts	Handicrafts[b]	Total	Population (in millions)	Per-capita output (in dong)	U. S. equivalent[c]	Total (in millions of U. S. $)	Per capita (U. S. $)
1955	1,565	317	225	1,882	13.5	139	533	533	39
1956	1,843	627	366	2,470	14.0	176	700	700	50
1957	1,894	865	496	2,759	14.5	190	782	782	54
1958	2,201	1,013	527	3,214	15.0	214	910	910	61
1959	2,384	1,311	596	3,695	15.5	238	1,047	1,047	68
1960	2,046	1,471	630	3,517	15.9	221	996	996	63
1961	2,225	1,671	679	3,896	16.5	236	1,104	1,104	67
1962	2,349	2,102	729	4,451	17.1	260	1,261	1,261	74
1963	2,309	2,261	747	4,570	17.7	258	1,295	1,295	73
1964	2,406	2,611	783	5,017	18.3	274	1,421	1,421	78
1965	2,507	2,706	812	5,213	18.9	276	1,477	1,477	78
1966	2,487	2,164	649	4,651	19.5	238	1,318	1,318	68
1967	2,467	1,731	519	4,198	20.0	210	1,189	1,189	59
1968	2,500	1,385	416	3,885	20.6	189	1,101	1,101	53
1969	2,605	1,468	440	4,073	21.2	192	1,154	1,154	54
1970	2,714	1,879	564	4,593	21.8	211	1,301	1,301	60

[a]Including some nonagricultural output produced by the rural sector.

[b]For 1966-70, value estimated as 30 percent of total value of industry and handicrafts.

[c]Converted at exchange rate of U. S. $1 = 3.53 dong.

Note: GMP is different from GNP in that service sectors are excluded and intermediate products are included. For 1955-60, GMP is in 1956 prices and for 1961-70, in 1959 prices; however, the difference between the two prices is insignificant because of strict price control and in light of the very small share of monetary transactions in total exchanges.

Sources: Estimates based on: DRVN, Central Statistical Office, Nam Nam Xay Dung Kinh Te va Van Hoa (Five years of economic and cultural building (Hanoi: Central Statistical Office, 1960), pp. 114, 158; So Lieu Thong Ke (Statistical data), 1960-63 (Hanoi, 1963), pp. 43, 57; Vo Nhan Tri, Croissance Economique de la Republique Democratique du Vietnam (1945-65) (Hanoi: Editions En Langues Etrangeres, 1967), pp. 229, 316-17, 357, 439-47; Hoc Tap, no. 5, 1961, p. 29; Leon Lavallee, L'Economie du Nord Vietnam (Paris: Centre d'Etude et de Recherche Marxiste, 1971), pp. 43, 90, 117; Le Chau, Le Viet-Nam Socialiste (Paris: Francois Maspero, 1966), pp. 294, 314, 328-29; William Kaye, "A Bowl of Rice Divided," in North Vietnam Today, ed. P. J. Honey (New York: Frederick A. Praeger, 1962), p. 114; Bui Quy Dap, "Nhung Bien Dei Cach Mang Sau Sac o Nong Thon va Trong Nong Nghiep Mien Bac" (Revolutionary and profound changes in North Vietnam's rural area and agriculture), Khoa Hoc Va Ky Tuat Nong Nghiep (Scientific and agricultural technology), no. 2, 1970, pp. 73-75; Jon M. Van Dyke, North Vietnam's Strategy for Survival (Palo Alto: Pacific Books, 1972), p. 109; Henri Dumesnil, "La Guerre et l'Economie de la Republique Democratique du Viet-Nam," Tiers Monde 11, no. 42-43 (August-September, 1970): 527-35; Le Monde, December, 23-25, 1970.

of time rather than a precise measure of the magnitude of national output in a given year. The estimates for aggregate economic trends in this chapter were calculated by (1) reconstructing an index of aggregate physical output by combining indexes of agricultural and industrial output for a number of years where data are not available; (2) using absolute value of a base year, converting these index series into national accounts; and (3) using growth rates of different sectors implied in different economic and political reports. These growth rates are, however, often inconsistent and misleading. Consequently, figures used in this chapter are rough approximations and tentative.

Gross Material Product

In the Soviet-type economy, GMP or gross social product (sovokupnyi obshchestvennyi produkt) (GSP) comprises material values produced by the country during one accounting period. In terms of value, GMP includes newly created value (national income), material inputs, and depreciation. Production in the Marxist tradition is defined as the product of social labor in material form.[2] As an important component of the service sector (the unproductive services) is excluded from the GMP the two most important sectors remaining are industry and agriculture including forestry and fishery. The sectors outside agriculture and industry were treated by North Vietnam's authority as playing a supporting role to the two main sectors. It is also possible that an important part of transportation services is already included in the calculation of agriculture and industry.

Taking these two sectors as representative of GMP, Table 6.1 may be taken as an indicator of the general trend of the economy during 1955 to 1970. Official data on the shares of different sectors in GMP are shown in Table 6.2.

The GMP grew rapidly during the period of reconstruction (1955-57) at an annual average rate of 23 percent, the highest in the country's economic history. Most of the increase took place in 1956 as both the agriculture and handicraft sectors had recovered from the 1945-54 war; the two sectors grew at 18 percent and 63 percent, respectively.

During the three-year plan period, growth slowed down to 9 percent per annum. Though still high, the rate could have been higher if it had not been for the decline in 1960 (5 percent) due to a sharp reduction in agricultural output because of adverse weather conditions in that year. The year 1959 was considered one of the best in North Vietnam since 1955 because of the excellent crop (5.7 million tons of paddy). A large increase in industrial output

TABLE 6.2

Shares of Different Sectors in
Gross Material Product
(in percentage)

Sector	1957	1960	1965	1968	1971
Industry and handicrafts	24.3	33.5	40.5	37.9	43.9
Agriculture	41.3	34.1	29.9	32.2	30.5
Construction	7.3	9.6	10.8	10.6	11.6
Transport and communication	2.4	2.7	2.6	3.3	3.3
Commerce and services	15.6	13.1	10.3	9.8	5.6
Others	9.1	7.0	5.9	6.2	5.1
Total	100.0	100.0	100.0	100.0	100.0

Source: DRVN, La Republique Democratique du Vietnam (Hanoi: Editions En Langues Etrangeres, 1975), p. 69.

due to substantial public investment in the modern sector during the first and second year of the plan was also a contributing factor to growth performance in 1959 (15 percent).

From 1961 to 1965, under the first five-year plan, GMP grew a total of 48 percent or 9.6 percent annually. High growth rates were recorded in 1961, 1962, and 1964 (11 percent, 14 percent, and 10 percent, respectively). As shown in the previous chapter, the plan accorded high priority to industry, with the consequence of raising the value of industrial output to 2.70 billion dong in 1965, compared to 1.47 billion dong in 1960—an average increase of 16.8 percent a year. The growth performance in agriculture was sluggish as the value of output in 1965 was 13 percent higher than the 1961 level and only 5 percent over 1959.

From an estimate of 5.21 billion dong in 1965, GMP dropped sharply to 4.65 billion dong in 1966, or by 11 percent, as the value of industrial production declined by 20 percent. The air war that began in late 1965 increased in intensity until early 1968 when it temporarily stopped. GMP growth rate continued to decline by 10 percent in 1967 and 7 percent in 1968, then increased by 5 percent and 13 percent, respectively, in 1969 and 1970 as the economy began to recover from the war. As national priority shifted from production to war, growth for the period of 1965-70 was negative and the 1970 level was 12 percent less than the 1965 level.

Trend lines for the GMP together with trends for agriculture and for industry and handicrafts for the 15-year period (1955-70) are presented along with their graphs in Figure 6.1. Their equations are as follows, with AGR indicating agriculture, INT, industry and handicrafts, and T, time:

$$\underset{}{GMP} = \underset{(325.7)}{2612.70} + \underset{(33.7)}{149.12}T \qquad (6.1)$$

$$AGR = \underset{(71.27)}{1784.75} + \underset{(7.37)}{58.43}T \qquad (6.2)$$

$$INT = \underset{(287.14)}{828.00} + \underset{(29.70)}{90.69}T \qquad (6.3)$$

Equation 6.3 indicates that industrial production increased faster than that of agriculture and was responsible for 60 percent of GMP increase during the period. The coefficients of T represent the average amount of increase per year for each associated dependent variable.

The relationship between these three aggregates derived from regression analysis is as follows:

$$GMP = \underset{(945.07)}{-2055.40} + \underset{(0.41)}{2.60}\ AGR \qquad R^2 = 0.74 \qquad (6.4)$$

$$INT = \underset{(945.07)}{-2055.40} + \underset{(0.41)}{1.60}\ AGR \qquad R^2 = 0.52 \qquad (6.5)$$

The coefficents of AGR represent the marginal contribution of AGR to the dependent variables; the values of parenthesis are standard error of coefficents. As evident in these equations, there is important and positive correlation between GMP and agricultural production and between agriculture and industry, the former correlation being the stronger. The scatters and the line of regression of these relationships are presented in Figure 6.2.)

With a population growing at an average of 3.5 percent a year, per-capita output in North Vietnam during the entire period of 15 years was among the lowest in the underdeveloped countries. With the exception of the three years 1962, 1964, and 1965, per-capita output was less than 260 dong per annum. Using an exchange rate

FIGURE 6.1

Trend Lines of Gross Material Product,
Agriculture, and Industry and Handicrafts, 1955-70
(at constant price and in millions of dong)

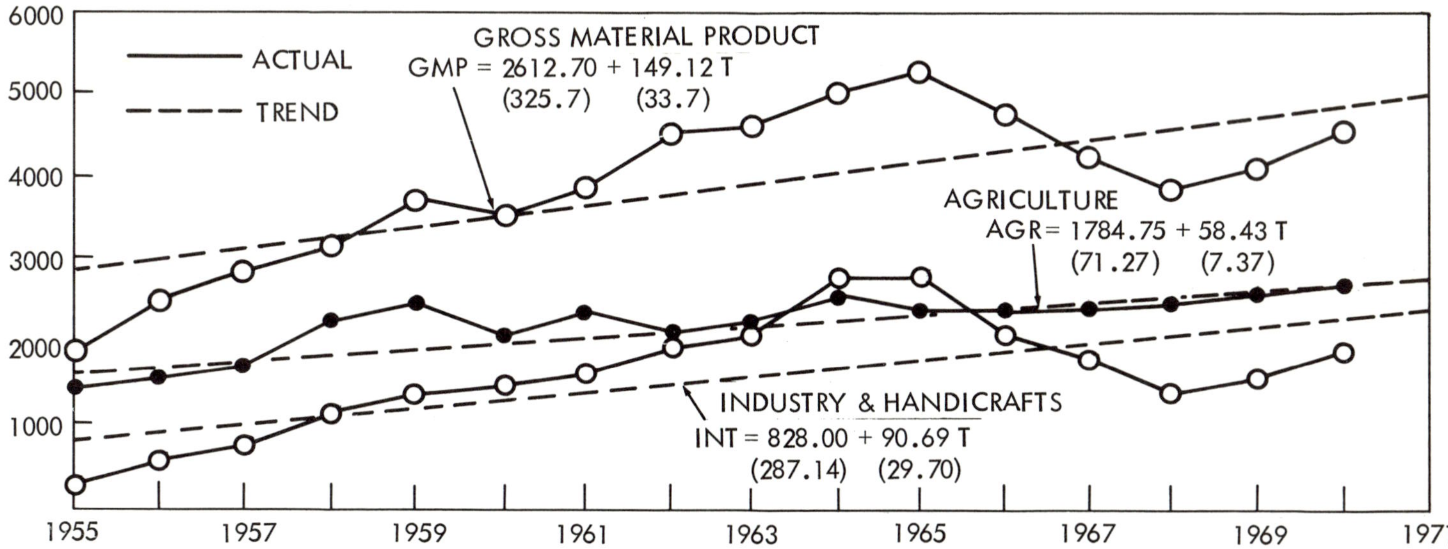

Sources: See page 114.

FIGURE 6. 2

Relationship between Gross Material Product,
Industry and Handicrafts, and Agriculture, 1955–70
(at constant million dong)

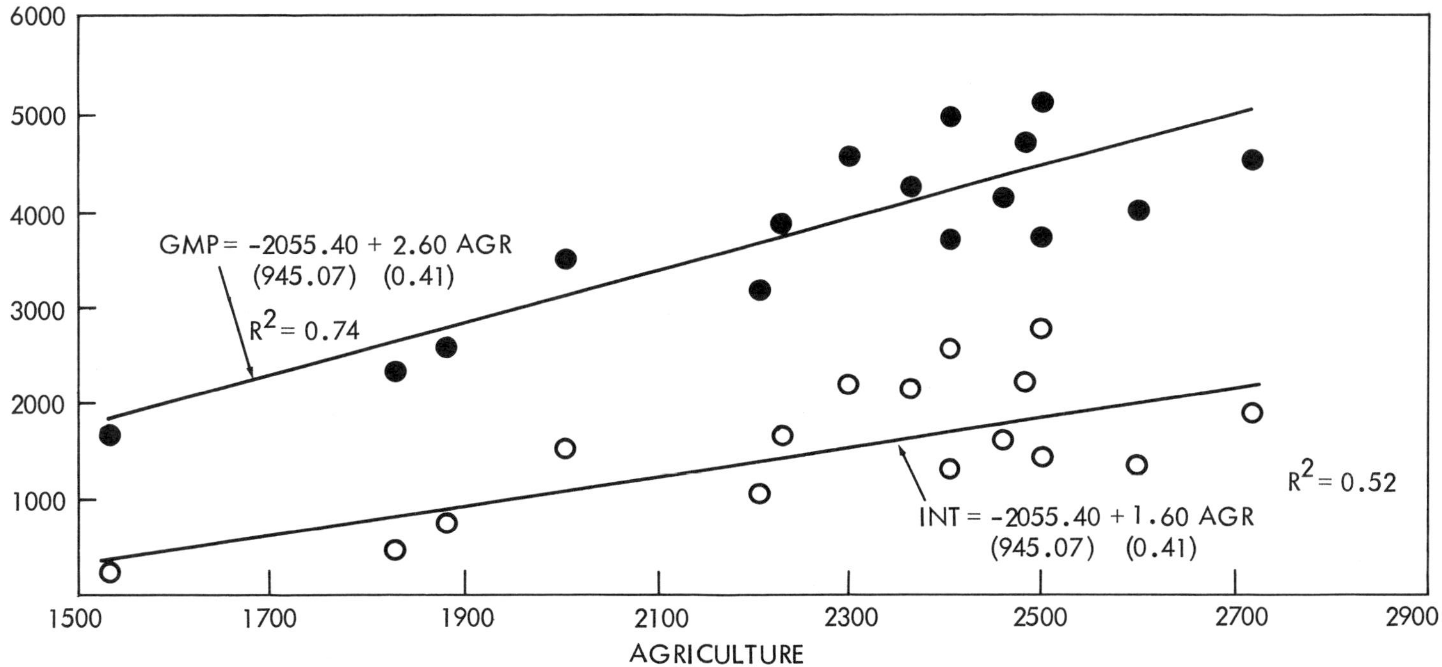

Sources: See Page 114.

TABLE 6.3

Shares of Different Sectors in National Income
(in percentage)

Sector	1957	1960	1965	1968	1971
Industry and handicrafts	16.0	18.2	23.1	19.9	28.3
Agriculture	44.4	42.3	39.8	43.9	42.1
Construction	3.4	4.7	6.8	6.4	7.9
Transport and communication	2.3	2.8	2.5	2.4	3.0
Commerce and services	20.8	20.5	17.3	16.4	8.9
Others	13.1	11.5	10.5	11.0	9.8
Total	100.0	100.0	100.0	100.0	100.0

Source: DRVN, La Republique Democratique du Vietnam, (Hanoi: Editions En Langues Etrangeres, 1975), p. 70.

of U.S. \$1 = 3.53 dong, to convert the figures for GMP, per-capita output was \$60 in 1970 compared with \$61 in 1958, \$68 in 1959, and \$63 in 1960. The highest level was for both 1964 and 1965.

The following equations, which are derived from Table 6.1, indicate that on the average, growth rates during the entire period of 1955-70 were 4.4 percent (least square estimated) for GMP, 8.1 percent for industry, and only 2.7 percent for agriculture. The values in parentheses are standard error of coefficients.

$$\text{Log GMP} = \underset{(0.095)}{7.851} + \underset{(0.009)}{0.044}T \qquad (6.6)$$

$$\text{Log INT} = \underset{(0.221)}{6.561} + \underset{(0.023)}{0.081}T \qquad (6.7)$$

$$\text{Log AGR} = \underset{(0.037)}{7.493} + \underset{(0.004)}{0.027}T \qquad (6.8)$$

TABLE 6.4

National Income and Per-Capita Income, 1957-65 (at constant prices)

Year	National Income[a]		Per-Capita Income[b]	
	Amount (millions of dong)	Growth Rate (percent)	Amount (dong)	Growth Rate (percent)
1957	2,624	--	181	--
1958	2,869	9	191	5.5
1959	3,352	17	216	13.1
1960	3,453	3	217	0.5
1961	3,694	7	224	3.2
1962	3,990	8	234	4.5
1963	4,070	2	231	-1.3
1964	4,395	8	241	4.3
1965	4,570	4	242	0.4

[a]Data on national income for 1960-65 are calculated on the basis of annual growth rate given by the Central Statistical Office.

[b]Per-capita income for 1961-64 is derived on the basis of estimated population.

Sources: Estimates based on: DRVN, Central Statistical Office, Nam Nam Xay Dung Kinh Te va Van Hoa (Five years of economic and cultural building) (Hanoi: Central Statistical Office, 1960), p. 32; Vo Nhan Tri, Croissance Economique de la Republique Democratique du Vietnam (1945-65) (Hanoi: Editions En Langues Etrangeres, 1967), p. 578; Leon Lavallee, L'Economie du Nord Vietnam (Paris: Centre d'Etude et de Recherche Marxiste, 1971), pp. 49, 117.

National Income

North Vietnam's Central Statistical Office published data on national income (thu nhap quoc dan) for 1957-59 and growth rates for 1961-63. Recently, data on shares of different sectors in national income have become available, and are shown in Table 6.3. Figures on national income presented in this chapter are compiled from official data for 1957-63 and from growth rates provided by various

North Vietnamese sources. It should be noted that the Soviet-type accounting system treats national income (natsional'nyi or narodnyi dokhod) differently from that used in the Western concept of national income. It is also referred to as net material product, which covers both income and output aggregates. National income, according to the Soviet concept, is the sum of labor earnings and surplus value,[3] its net-material-products equivalents measuring newly created value equal to GMP minus intermediate distribution of output including depreciation.[4]

Table 6. 4 shows that national income increased at an average rate of 10. 7 percent during 1957-60 with the highest rate achieved in 1959 (17 percent). The spectacular and unique growth rate in 1959 was due to the excellent performance of the agricultural sector, as already mentioned. During the subsequent six years, national income grew at a much less rapid rate, averaging 6 percent per annum with an annual rate as low as 2 percent in 1963. In monetary terms, national income in 1965 is estimated at 4. 57 billion dong (U. S. $1. 3 billion) compared with 3. 45 billion dong in 1960 and 2. 62 billion dong (U. S. $743 million) in 1957.

Using the same set of estimates for population as shown in Table 6. 1, per-capita income, which was 181 dong (U. S. $51) in 1957, grew rapidly to 216 dong (U. S. $61) in 1959. Thereafter, progress was slow, with per-capita income estimated at only 242 dong (U. S. $69) for 1965. Per-capita income growth rate was highest in 1959 (13. 1 percent) and lowest in 1963 when a negative growth of 1. 3 percent was observed. The average rate for the entire 1957-65 period was 4. 25 percent.

Table 6. 5 shows the distribution, by sectors, incomes generated in the production of national income. The value of each sector in this table is smaller, and in the case of industry, significantly smaller, than the corresponding value in Table 6. 1 because Table 6. 1 measures value of gross social product as compared with net-social-product value measured in Table 6. 5.

Agriculture, which increased considerably during 1957-60, at an annual average rate of 8 percent, dropped sharply during the following five years to an annual average rate of 2. 2 percent; its 1965 level was barely greater than the 1962 level. On the contrary, the relative position of the industry and handicrafts sector in the generation of national income increased rapidly during the 1960-65 period (17. 8 percent per annum) as compared with the 1957-60 period (12. 7 percent per annum). As a result, the composition of national income underwent a structural change, which reflects North Vietnam's heavy emphasis on development of industry. Industry's share in total national income was only 17. 3 percent in 1957 and increased to 26 percent in 1965, while agriculture's share was reduced from 57

TABLE 6.5

Contributions to National Income, by Sector, 1957-65
(at constant prices and in millions of dong)

Sector	1957		1960		1962		1963		1964		1965[b]	
	Amount	Percent	Amount	Percent	Amount	Percent	Amount	Percent	Amount	Percent[a]	Amount	Percent
Agriculture	1,496	57.0	1,851	53.6	1,991	49.9	1,978	48.6	2,022	46.0	2,057	45.0
Industry and handicrafts	454	17.3	628	18.2	866	21.7	924	22.7	1,099	25.0	1,188	26.0
Infrastructure	66	2.5	162	4.7	259	6.5	256	6.3	264	6.0	274	6.0
Transport and communications	52	2.0	97	2.8	112	2.8	126	3.1	175	4.0	183	4.0
Commerce	556	21.2	715	20.7	762	19.1	786	19.3	835	19.0	868	19.0
Total	2,624	100	3,453	100	3,990	100	4,070	100	4,395	100	4,570	100

[a]Percentage shares are estimated.

[b]Estimated.

Note: Individual items may not add to totals because of rounding.

Sources: Estimates based on: DRVN, Central Statistical Office, Nam Nam Xay Dung Kinh Te va Van Hoa (Five years of economic and cultural building) (Hanoi: Central Statistical Office, 1960), p. 32; Vo Nhan Tri, Croissance Economique de la Republique Democratique du Vietnam (1945-65) (Hanoi: Editions En Langues Etrangeres, 1967), p. 578; Leon Lavallee, L'Economie du Nord Vietnam (Paris: Centre d'Etude et de Recherche Marxiste, 1971), pp. 49, 117.

TABLE 6.6

Expenditures on National Income, 1957-65
(at constant prices and in millions of dong)

Year	Consumption	Expenditure Consumption as Percentage of National Income	Investment	Investment as Percentage of National Income
1957	2,335	89.0	289	11.0
1958	2,453	85.5	416	14.5
1959	2,705	80.7	647	19.3
1960	2,818	81.6	635	18.4
1961	3,014	81.6	680	18.4
1962	3,300	82.7	690	17.3
1963	3,382	83.1	688	16.9
1964	3,648	83.0	747	17.0
1965[a]	3,793	83.0	777	17.0

[a]Estimated.

Sources: Estimates based on: DRVN, Central Statistical Office, Nam Nam Xay Dung Kinh Te va Van Hoa (Five years of economic and cultural building (Hanoi: Central Statistical Office, 1960), pp. 114, 158; So Lieu Thong Ke (Statistical data), 1960-63 (Hanoi, 1963), pp. 43, 57; Vo Nhan Tri, Croissance Economique de la Republique Democratique du Vietnam (1945-65) (Hanoi: Editions En Langues Etrangeres, 1967), pp. 229, 316-17, 357, 439-47; Hoc Tap, no. 5, 1961, p. 29; Leon Lavallee, L'Economie du Nord Vietnam (Paris: Centre d'Etude et de Recherche Marxiste, 1971), pp. 43, 90, 117; Le Chau, Le Viet-Nam Socialiste (Paris: Francois Maspero, 1966), pp. 294, 314, 328-29; William Kaye, "A Bowl of Rice Divided," in North Vietnam Today, ed. P.J. Honey (New York: Frederick A. Praeger, 1962), p. 114; Bui Quy Dap, "Nhung Bien Dei Cach Mang Sau Sac o Nong Thon va Trong Nong Nghiep Mien Bac" (Revolutionary and profound changes in North Vietnam's rural area and agriculture), Khoa Hoc Va Ky Tuat Nong Nghiep (Scientific and Agricultural technology), no. 2, 1970, pp. 73-75; Jon M. Van Dyke, North Vietnam's Strategy for Survival (Palo Alto: Pacific Books, 1972), p. 109; Henri Dumesnil, "La Guerre et l'Economie de la Republique Democratique du Viet-Nam," Tiers Monde 11, no. 42-43 (August-September, 1970): 527-35; Le Monde, December 23-25, 1970.

percent to 45 percent. Under the impact of the intensification of the war after 1965, however, the trend appears to have been reversed, with the relative position of agriculture becoming increasingly important as major sectors of industry were either disrupted or decreased in output. Among other sectors, both infrastructure, and transport and communications, though relatively small, grew rapidly during the 1957-65 period. Their shares in national income were 6 percent and 4 percent, respectively, in 1965 compared to 2.5 percent and 2 percent, respectively, in 1957. The role of commerce in national income declined slightly from 21.2 percent in 1957 to 19 percent in 1965.

Consumption and Investment

Breakdown of national income into expenditures on consumption and investment is given in Table 6.6, which is derived from figures on national income as estimated in Table 6.4 and from the percentage share of each category of expenditures given by the North Vietnamese sources. Again, according to the socialist accounting system, the treatment of consumption and investment is different from the Western concepts.[5]

Consumption expenditure, which was 89 percent of national income in 1957, was reduced sharply to 80.7 percent in 1959, then rose slowly to 83 percent in the final year of the first five-year plan. The consequence of the reduction in consumption is the relative increase in investment, which averaged 15 percent of national income during 1957-59 and 17.5 percent during 1960-65.

Trend lines fitted to national income (NI), consumption (C), and investment (INV), for 1957-65 are shown in Figure 6.3, and their equations are:

$$NI = \underset{(65.9)}{2474.0} + \underset{(11.7)}{238.9}T \qquad (6.9)$$

$$C = \underset{(32.1)}{2112.0} + \underset{(5.7)}{187.5}T \qquad (6.10)$$

$$INV = \underset{(59.9)}{361.9} + \underset{(10.6)}{51.4}T \qquad (6.11)$$

The relationships between consumption, investment, and national income are shown in Figure 6.4, and their equations are as follows:

FIGURE 6.3

Trend Lines for National Income, Consumption, and Investment, 1957-65
(at constant million dong)

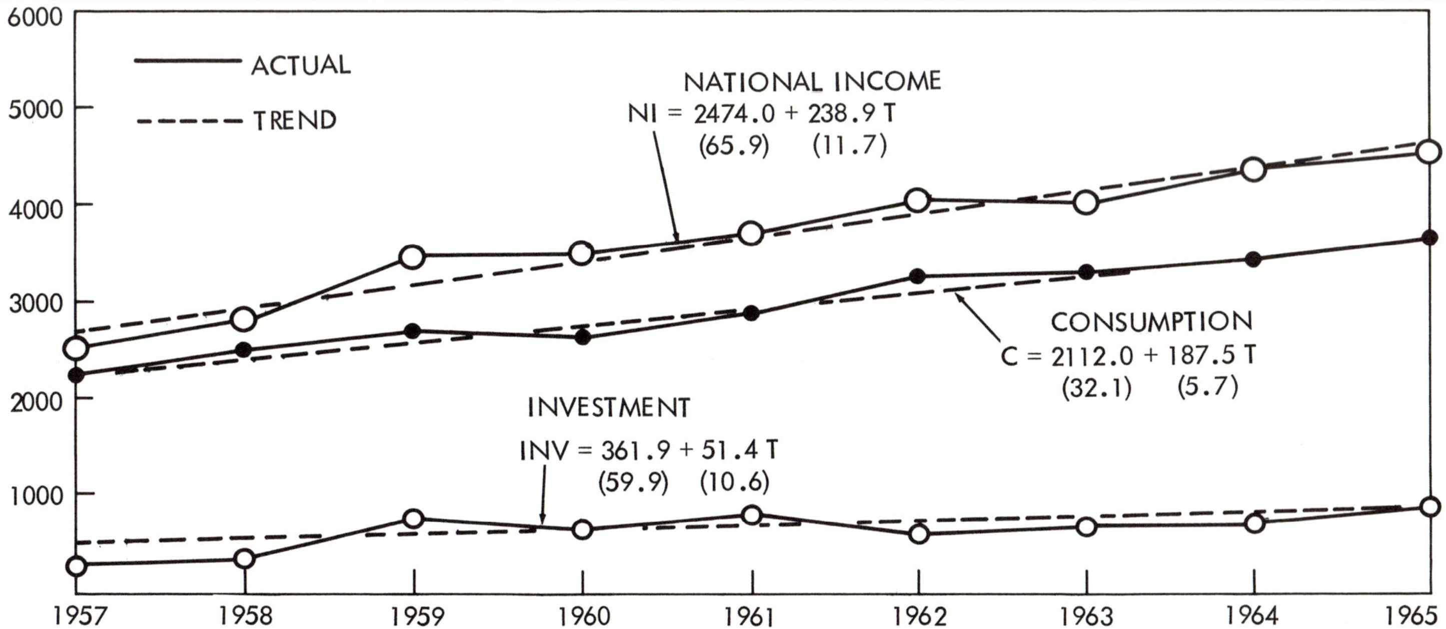

Sources: See page 114.

FIGURE 6.4

Relationship between Consumption and National Income, and between Investment and National Income, 1957-65 (at constant million dong)

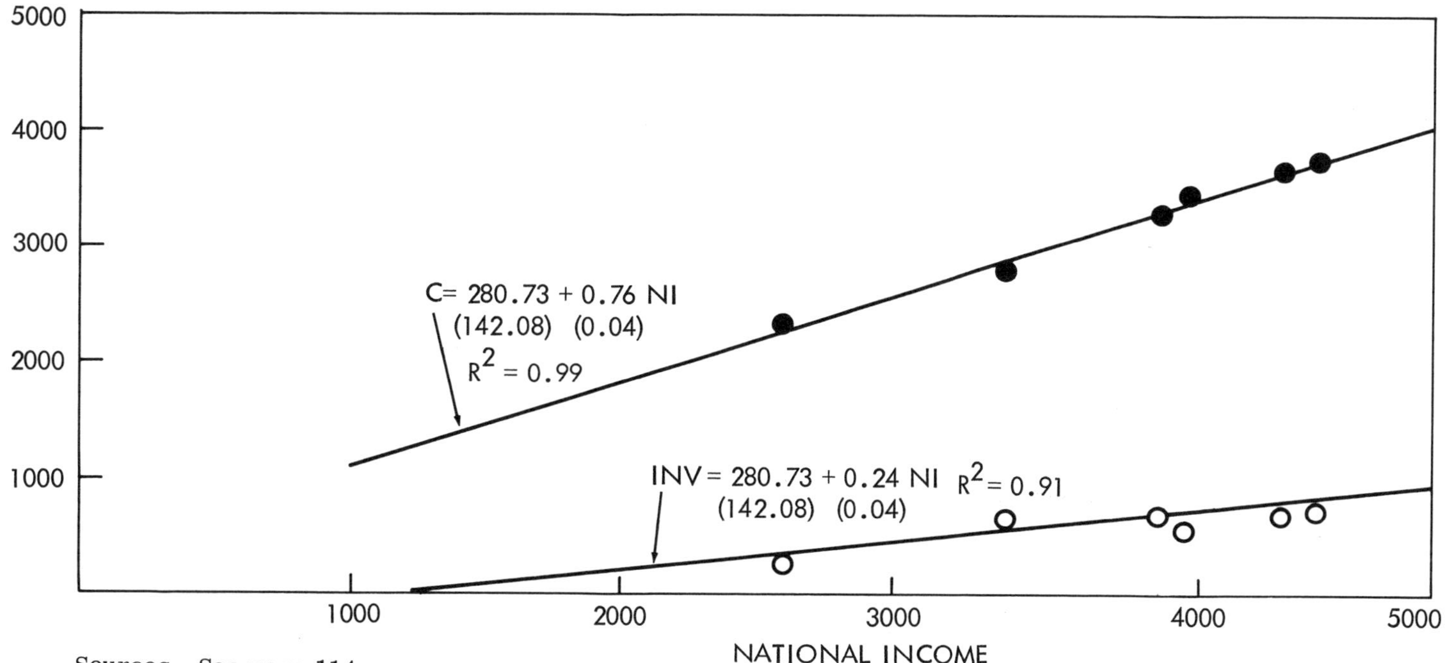

Sources: See page 114.

$$C = \underset{(142.08)}{280.73} + \underset{(0.04)}{0.76NI} \qquad R^2 = 0.99 \tag{6.12}$$

$$INV = \underset{(142.08)}{280.73} + \underset{(0.04)}{0.24NI} \qquad R^2 = 0.91 \tag{6.13}$$

Equation 6.12 is a consumption function, which indicates that marginal propensity to consume was 0.76 during the period. The latter implies a high marginal propensity to save of 0.24. The following equations show that over the eight years between 1957 and 1965, the unique period of relative economic progress in North Vietnam, growth rates averaged 6.7 percent a year for national income, 6.2 percent for consumption, 9.8 percent for investment, and 3.4 percent for per-capita income (PI). However, because of the conceptual difference between the Western and socialist measurement of national income, the per-capita income calculated here does not reflect the improvement in the welfare of people. Furthermore, because of substantial production decline since 1965, per-capita income growth rate has been much smaller.

$$LogNi = \underset{(0.028)}{7.856} + \underset{(0.005)}{0.067T} \tag{6.14}$$

$$Log\ C = \underset{(0.012)}{7.698} + \underset{(0.002)}{0.062T} \tag{6.15}$$

$$LogINV = \underset{(0.140)}{5.89} + \underset{(0.024)}{0.098T} \tag{6.16}$$

$$LogPI = \underset{(0.028)}{5.215} + \underset{(0.005)}{0.034T} \tag{6.17}$$

CONCLUSION

Quantitative trends analyzed in this chapter show that, common to the situation in other economies under reconstruction, GMP growth in the North Vietnamese economy was impressive during 1955-59 (23 percent on average). Growth during the two plans that followed (1958-65), though substantially less impressive, was still relatively high (over 9 percent as an annual average). During the latter half of the 1960s, however, the economy regressed under warfare, and growth rate became negative. Averaging out, the overall GMP growth was 4.4 percent per annum during the 1955-70 period. Because of national priority given to industry, average industrial growth during the same period (8.1 percent) was almost twice as large as that of GMP with most of the increase taking place in the first five-year plan. The high rate of industrial expansion was contrary to the slow growth of agriculture, 2.7 percent per annum, a rate which is 0.8 percent behind the population growth rate.

Estimates for NI show that, in the main, growth performance followed the same trend as that of GMP. The difference was that the NI growth rate, which was only half that of GMP during reconstruction (10.7 percent on average against 23 percent) drew closer to GMP during the two plans that followed (6 percent compared to 9 percent). Also during the two plans, the marginal propensity to consume is estimated at 0.76 percent, implying a relatively high marginal propensity to save (0.24). As a result, investment grew at an annual average rate of nearly 10 percent during the same period.

In monetary terms, per-capita income in 1965 was estimated at 242 dong (U.S. $69) as compared to 181 dong (U.S. $51) in 1957, an increase of 73 percent during the eight years of economic development under the plans. North Vietnam's per capita income, even at the height of development (1957-65), was indeed among the lowest in the world.

NOTES

1. For a thorough comparison of the Western and Soviet types of national accounting systems, see Morris Bornstein, "A Comparison of Soviet and United States National Product," in The Soviet Economy, ed. Harry G. Shaffer (New York: Appleton-Century-Crofts, 1963), pp. 58-69; V. N. Starovskii, "On the Methodology of Comparing Economic Indices of the U.S.S.R. and the U.S.A.," in ibid., pp. 46-56; Abraham Becker, "National Income Accounting in the U.S.S.R.," in Soviet Economic Statistics, ed. Vladimir G. Treml and John P. Hardt (Durham, N.C.: Duke University Press, 1972), pp. 69-119.

2. The labor is social when it is regular and gainful and when the work is reproducible. For example, occasional fishing for pleasure is not social labor; a piece of art also is not social labor. The product is material if it involves a physical good. See Becker, op. cit. pp. 71-73.

3. Labor earnings include earnings of productive labor in money and in kind, including earnings from self employment (for example, income from private land). Surplus value included items such as collective farm deductions from social funds, social-security taxes, profits. See ibid., p. 86.

4. The concept of depreciation is also different from Western concepts. Depreciation on nonproductive capital (housing, office

equipment, structures) is considered a consumption charge on output (ibid., pp. 76-77).

5. See ibid., pp. 89-91; and Stanley H. Cohn, "National Income Growth Statistics," in Soviet Economic Statistics, ed. Treml and Hart, pp. 140-47.

FIGURE SOURCE NOTE

Estimates based on: DRVN, Central Statistical Office, Nam Nam Xay Dung Kinh Te va Van Hoa (Five years of economic and cultural building) (Hanoi: Central Statistical Office, 1960), pp. 114, 158; So Lieu Thong Ke (Statistical data), 1960-63 (Hanoi, 1963), pp. 43, 57; Vo Nhan Tri, Croissance Economique de la Republique Democratique du Vietnam (1945-65) (Hanoi: Editions En Langues Etrangeres, 1967), pp. 229, 316-17, 357, 439-47; Hoc Tap, no. 5, 1961, p. 29; Leon Lavallee, L'Economie du Nord Vietnam (Paris: Centre d'Etude et de Recherche Marxiste, 1971), pp. 43, 90, 117; Le Chau, Le Viet-Nam Socialiste (Paris: Francois Maspero, 1966), pp. 294, 314, 328-29; William Kaye, "A Bowl of Rice Divided," in North Vietnam Today, ed. P. J. Honey (New York: Praeger, 1962), p. 114; Bui Quy Dap, "Nhung Bien Dei Cach Mang Sau Sac o Nong Thon va Trong Nong Nghiep Mien Bac" (Revolutionary and profound changes in North Vietnam's rural area and agriculture), Khoa Hoc Va Ky Tuat Nong Nghiep (Scientific and agricultural technology), no. 2, 1970, pp. 73-75; Jon M. Van Dyke, North Vietnam's Strategy for Survival (Palo Alto: Pacific Books, 1972), p. 109; Henri Dumesnil, "La Guerre et l'Economie de la Republique Democratique du Viet-Nam," Tiers Monde 11, nos. 42-43 (August-September, 1970): 527-35; Le Monde, December 23-25, 1970.

PART IV
ECONOMIC PERFORMANCE

CHAPTER

7

THE AGRICULTURAL SECTOR

BACKGROUND

Ever since taking power in 1954, the Communist regime has been determined to pursue one overriding goal: the transformation of an agricultural economy into a socialist industrial state. However, as North Vietnam entered the second development decade, agriculture still remained the predominant sector in its economy. In 1971 agriculture's share in national income remained roughly at the same level as in 1965 (42 percent), and slightly smaller than in 1957 (44 percent). In 1971 agricultural activities still provided employment for nearly 77 percent of the labor force as against 84 percent in 1965 and 85 percent in 1960.

One of the basic problems in North Vietnam's agriculture involves the irregularity of rainfall, the vulnerability of the dikes, and the violent character of the Red River, as already noted in Chapter 1. These natural conditions have greatly offset the advantage otherwise experienced in a tropical climate that permits year-round cultivation, especially in the case of rice. Given an adequate and timely supply of water, rice can be produced almost throughout the year in the tropics because of the length of the day, which is never too short even in the winter, in contrast to the higher latitudes.[1] But the key problem in rice cultivation is not only the availability of water, but, more important; the amount of it. North Vietnam, however, has not succeeded in controlling the water of the Red River, and as recently as 1975, some 300,000 hectares (12 percent of the area under food-crop cultivation) were submerged.[2]

The rugged topography, soil conditions, and climate have restricted the cultivation of crops to only slightly over 12 percent of

TABLE 7.1

Area under Cultivation, 1955-73
(thousand hectares)

Year	Food grains			Fruits and Vegetables	Industrial	Total
	Rice	Subsidiary	Total			
1955	2,176	391	2,567	15	69	2,651
1957	2,192	353	2,545	16	105	2,666
1958	2,235	367	2,602	26	114	2,742
1959	2,274	380	2,654	29	117	2,800
1960	2,284	420	2,704	30	137	2,871
1961	2,410	552	2,962	36	152	3,150
1962	2,410	587	2,997	51	186	3,234
1963	2,363	639	3,002	39	196	3,237
1964	2,431	690	3,121	121	207	3,449
1961-64 Average	2,404	617	3,021	117	200	3,338
1965-68 Average	2,269	676	2,945	142	207	3,294
1969-71 Average	2,149	570	2,719	152	177	3,048
1972	2,200	555	2,755	158	186	3,099
1973	2,094	519	2,613	156	167	2,936

Sources: DRVN, Etudes Vietnamiennes, no. 2, 1964, p. 178, no. 13, 1967, pp. 7-21, no. 44, 1976, pp. 171-78.

total land area (as compared with 50 percent forest land).[3] Most of the land that is not already tapped is in marginal areas in the mountainous region, where altitude, aridity, and diseases (especially malaria) discourage farming.

About 75 percent of the population and a similar percentage of cultivable land are concentrated in the Red River delta.[4] In order to expand land under cultivation, the government has launched campaign after campaign including forced transfer of people from the coastal plain to open new land in the highland. The campaigns, however, have had very limited results simply because there is not much cultivable land left.

Thus, with virtually no new arable land, North Vietnam has also turned to improving land productivity by increasing multiple-crop areas, introducing new crops and better strains of seeds, raising the level of agricultural technology, and more intensive application of inputs. Nevertheless, production remains stagnant. In addition to the problems of weather and topography, other factors contributing to poor performance in agriculture are lack of incentives (which is often referred to as the "management problem"), labor shortage (because of the drain of manpower war activities), low capital investment, and low level of technology.

LAND UTILIZATION

The structure of crop land is presented in Tables 7.1 and 7.2. The area under rice cultivation amounted to only 2.2 million hectares in 1955, implying a man-land ratio of 6.2 persons per hectare. In spite of efforts to recover cultivable land abandoned during the 1945-55 war and to open new land, a limit in rice land of around 2.4 million hectares, with year-to-year fluctuations due to weather conditions, appears to have been reached in the early 1960s. This level declined sharply in the late 1960s and early 1970s, and the 1972-75 level corresponds roughly to the 1955 figure, partly because of the agricultural-diversification policy and partly because of natural calamities, especially during 1971, 1973, and 1975.

With rice land in 1973 estimated at 2.1 million hectares and with a population estimated at 24.8 million in that year, the implicit man-land ratio was nearly 12 persons per hectare, or almost twice as high as the 1955 ratio.

To supplement the traditional rice diet, the government has increased acreage under cultivation in subsidiary crops and encouraged the population to consume more sweet potatoes, corn, and manioc. The area under subsidiary-crop cultivation increased sharply from 391,000 hectares in 1955 to 690,000 hectares in 1964, or by 76 percent, pushing the total area under food-grain production to a

TABLE 7. 2

North Vietnam: Pattern of Land Utilization, 1955-73
(average, in percentage)

Crop	1955	1960	1961-64	1965-68	1969-71	1972	1973
Food	97.0	94.0	90.5	89.4	89.2	88.9	89.0
Rice	82.0	80.0	72.0	70.0	70.0	71.0	71.0
Fruits and vegetables	1.0	1.0	3.5	4.3	5.0	5.1	5.3
Industrial	3.0	5.0	6.0	6.3	5.8	6.0	5.7
Total	100.0	100.0	100.0	100.0	100.0	100.0	100.0

Note: Figures may not add up to totals because of rounding.
Sources: DRVN, Etudes Vietnamiennes, no. 13, 1967, p. 20, no. 44, 1976, p. 177.

record high of over 3.1 million hectares, an increase of over half a million hectares compared to 1955. The expansion of area was assisted by a transfer of nearly 700,000 people from the coastal plains to the highland.

In addition to grain production, an effort has also been made to increase the growing of vegetables and fruit trees. Areas under fruit and vegetable cultivation increased from around 15,000 hectares in 1955 to double that amount in 1960, and to 156,000 hectares in 1973, a more than tenfold increase. The increase took place mainly in the suburbs of large urban centers such as Hanoi and Haiphong where cooperatives grow fruits and vegetables on large plantations.

An important feature of North Vietnam's land utilization is the expansion of area for industrial crops such as cotton, soybeans, groundnuts, sugarcane, tea, and coffee. Industrial-crop land, which averaged around 200,000 hectares during 1961-64, increased to 207,000 hectares during 1965-68, which represented a threefold increase over 1955. In the 1970s, however, it declined, and the 1972 estimated data remain the same as that of a decade earlier.

Table 7.2 shows the overall picture of the changing pattern of land utilization. Food crops, which shared 97 percent of total cultivated land in 1955, declined to 89 percent in 1973, with rice crops accounting for all the decline since land under subsidiary-crop cultivation actually increased (from 15 to 18 percent). During the same period, the area devoted to fruits and vegetables increased fastest, from 1 to 5 percent of the total, while industrial-crop area increased from 3 to 6 percent.

Prior to the unification of the country in 1976, the North Vietnamese government's ambitious land plan has sought to increase agricultural land to 29 percent of total land by 1985 (against the present 12-13 percent) confining land expansion to the midland and highland region. In the same year, according to the plan, land with agricultural capacity would reach 5 percent of total land, forest land would remain at 50 percent, and other cultivation would account for 16 percent.[5]

AGRICULTURAL POLICIES

Given the climate and topographical conditions in North Vietnam (as already noted), the goal of more than doubling agricultural land by 1985 appears to be illusory. It seems rather clear that any appreciable increase in food production in North Vietnam must come from factors other than expansion of arable land.

Until the present, agricultural policies have consisted of three main components: establishment of new relations in production, technical improvement in agriculture, and "cultural and ideological revolution" in the countryside. The first and the third sets of policies may be linked together and translated into two specific policies: land reform and collectivization. Both policies are supported by indoctrination of the masses following the Marxist-Leninist ideology.[6] As the main thrust of these policies was discussed in Chapter 4, only the technical measures are dealt with in this section.

At the heart of North Vietnam's agricultural policy is the emphasis on intensive cultivation in order to increase land productivity to an ambitious goal of 5 tons per hectare. Toward this goal, agricultural policy has focused on multiple cropping to raise the rate of land utilization, and on irrigation, fertilizer, and mechanization to increase productivity.

Multiple Cropping

Traditionally, Vietnamese farmers grow two crops per year. One is the winter crop (vu chiem), with sowing in October-November (cold and dry season) and harvesting in May-June (which is why it is also called the fifth-month crop). The other is the autumn crop (vu mua), the main crop, with sowing in May-June (hot and rainy season) and harvesting in October-November (it is also called the tenth-month crop).

Since 1965 two more crops have been introduced: the spring crop, with sowing in February-March and harvest in June; and the summer crop, with sowing in April-May and harvest in July-August. The spring crop (introduced to take advantage of the weather conditions in the dry season) has received increasing emphasis due to the introduction of new rice varieties: the Tran Chau in 1966-67 and the Nong Nghiep 8 in 1969.[7] Of total area under cultivation during the dry season, the spring-crop area, which amounted to only 6.3 percent in 1968, increased sharply to 58 percent in 1971, thus surpassing the area of the traditional dry-season crop (the fifth-month crop).[8] The spring-crop area was expected to increase from 540,000 hectares in 1971[9] to 800,00 hectares in 1975.[10]

The goal of multiple cropping includes the cultivation of 400,000 hectares of two rice crops and one subsidiary crop (corn or sweet potatoes) instead of only two rice crops, 600,000 hectares of two rice crops plus one azolla crop,[11] and 100,000 hectares of two or three subsidiary crops.

Water Supply

Water represents the most difficult problem in North Vietnam's agriculture (as already discussed). Policies to solve the problem include first, and above all, defensive measures to prevent the crops from being destroyed and, second, dynamic measures to increase yield. High-yield varieties of rice require not only availability of water but also the control of water level at different stages of growth of the rice plant. Fully aware of the role of water in the rice economy, the regime has accorded water highest priority in the state's agricultural investment. Table 7.3 shows that of total budgetary investment in agriculture, 84 percent, 53 percent, and 54 percent were allocated for irrigation and water control during the reconstruction plan (1955-57), the three year plan (1958-60) and the five-year plan (1961-65), respectively. In 1971, the state's investment in this area is estimated to have increased to 62 percent of total investment in agriculture. Nevertheless, since investment in agriculture amounted to only between 12 percent (1958-60) to 21 percent (1961-71), investment in water was actually very modest. It is claimed that the ratio of area irrigated to total cultivated land reached 69 percent in 1968 as compared to 64 percent in 1960. In the context of Asian agriculture, North Vietnam's policy to reserve the very limited amount of agriculture investment for irrigation appears to be a rational one, since in the case of rice the leading input up to a certain stage of development is water, the proper

TABLE 7.3

Budgetary Investment in Agriculture, 1955-71
(in millions of dong)

Year	Total	Investment in Irrigation	Total As Percent of Budgetary Investment	Investment in Irrigation As Percent of Total
1955	23	21	17	91
1956	36	29	16	81
1957	28	22	13	79
1958	48	35		73
1959	57	25	12	44
1960	71	29		41
1961	109	49	16	45
1962	149	72	21	48
1963	163	79	23	48
1964	175	111	23	64
1965	164	108	...	66
1965-68 (average)	157	97	14	62
1969-71 (average)	231	147	21	64

Sources: For 1955-65: DRVN, So Lieu Thong Ke, 1955-59 (Hanoi, 1960); p. 99; Nghien Cuu Kinh Te, no. 39, 1967, p. 13. For 1965-71: DRVN, Etudes Vietnamiennes, no. 44, 1976, p. 196; DRVN, La Republique Democratique du Vietnam (Hanoi: Editions En Langues Etrangeres, 1975), pp. 67, 76-79.

supply and control of which plays a crucial role in determining output. On the basis of several studies on Southeast Asian agriculture, production in that stage of development is around 2.3 tons per hectare, which is about the average yield estimated for present-day North Vietnamese agriculture. In this stage, marginal increases in productivity seem to be most significantly explained by the irrigation ratio.[12]

Fertilizer

In addition to full exploitation of the traditional sources of fertilizer, the organic types,[13] fertilizer policy in recent years has focused mainly on the production of the so-called green fertilizer, especially the azolla type. Azolla, a nitrogen-rich green slime, has been used for several centuries, but was restricted to a small area in the coastal districts of the delta (especially in Thai Binh province).[14] In 1955 total area under azolla cultivation was around 40, 000 hectares. The government launched a campaign during 1965-68 to expand massively the area under azolla cultivation, which, as a result, reached 400, 000 hectares grown mainly in the winter after the harvesting of the tenth-month crop. As such, green fertilizer now supplies between 25 and 30 percent of total fertilizer requirement. The goal is to increase the area under azolla to 700, 000-800, 000 hectares in the coming years.[15]

As the Viet Tri chemical-fertilizer plant was destroyed during the air war, nearly all of the chemical fertilizer supplied by the state to cooperatives was imported. According to North Vietnam's source, import in 1965 had to increase by 196 percent over 1964 and in 1968, by 125 percent over 1967.[16]

Mechanization

As a result of the drain of manpower from the rural area to war-related activities, agricultural mechanization was started in 1965 to supplement the female labor force, which accounted for between 70 and 80 percent of workers in agriculture.[17] Mechanization took place with the introduction of small motors with 12 to 15 horsepower for pumping water, husking, and threshing. Side by side with mechanization, cooperatives were merged in order to increase the average size from 100 to 200 hectares to render economical the use of motors. It is estimated that the use of motors increases productivity of farmers to 20 times above that when working with buffaloes on the paddy fields;[18] the productivity of a mechanical thresher is three or five times that of hand- or buffalo-driven threshers, and the cost is 20 to 60 percent lower. Output of rice-husking machinery is 20 to 25 times more than that of hand-driven mills with a cost reduction of between one-third to one-fifth.[19]

By 1972, it is claimed that 52. 3 percent of all cooperatives possessed small machines as compared to only 2. 5 percent in 1965. The number of 15-horsepower motors increased sevenfold compared to 1964, while area under mechanized farming increased nearly fivefold. In the same year there was one 15-horsepower motor for every 394 hectares cultivated as against one for every 2, 750 hectares in 1965.[20]

TABLE 7.4

Contribution of Agricultural Activities to
Total Value of Agricultural Production, 1939-73
(in percentage)

Activity	1939	1955 -60	1961 -64	1965 -68	1969 -71	1972	1973
Cultivation	84	84	81	80	79	80	77
Food crops	60	66	60	55	53	—	—
Fruits and vegetables	19	15	16	25	20	—	—
Industrial crops	5	3	5	5	6	—	—
Animal husbandry and fishing	16	16	19	20	21	20	23

Sources: For 1955-65: DRVN, So Lieu Thong Ke, 1955-59 (Hanoi, 1960); p. 99; Nghien Cuu Kinh Te, no. 39, 1967, p. 13. For 1965-71: DRVN, Etudes Vietnamiennes, no. 44, 1976, p. 196; DRVN, La Republique Democratique du Vietnam (Hanoi: Editions En Langues Etrangeres, 1975), pp. 67, 76-79.

AGRICULTURAL PERFORMANCE

Agriculture includes two major activities: cultivation, and animal husbandry and fishing. Cultivation, which is composed of food crops, fruits and vegetables, and industrial crops, contributes the largest share in the value of agricultural production. Its share, however, has declined steadily in the 1960s in favor of animal husbandry. As shown in Table 7.4, the share of crop raising, which averaged 84 percent of total value during 1955-60, declined to 80 percent during 1965-68 and to 77 percent in 1973. At the same time, the value share of animal husbandry increased from 16 percent to 20 and 23 percent, respectively, or by 44 percent. The increasingly important role of animal husbandry is expected to continue as it is an important aspect of North Vietnam's long-term development policy.

Within the crop-raising sector, food-crop value share declined steadily to an estimated 53 percent for 1973 as compared to a 66 percent average during 1955-60, while fruit and vegetable production tended to rise especially during 1965-66 when its share in total agricultural value averaged 25 percent.

The production of food grains is shown in Table 7.5. Official statistics for rice and subsidiary-crop production are available for the period 1955-63. Only growth rates are given for later years. Figures for 1964-76, therefore, are reconstructed from various North Vietnamese sources (1964), from estimates based on growth rates (1965-73), or from the planned levels (1974-76).

Rice

At the dawn of the regime, an all-out effort was launched to "solve the question of food" and "win the battle of rice."[21] It was thought that land reform and collectivization could be the effective instruments to solve the food problem. In addition, priority was given to the restoration of the irrigation system and to the building of new hydraulic works involving large-scale state-investment spending and local labor. Since the period of reconstruction (1955-57), the government has turned to laying emphasis on socialist industrialization. Nevertheless, the agricultural problem continues to loom so big for the country that North Vietnam never seems to be able to reach the goal of 7 million tons of rice set for the terminal year of the first five-year plan (1961-65). Attainment of the dream of socialist industrialization is critically dependent upon the ability of agriculture to produce a surplus of food and raw materials above rural consumption to support labor engaged in industry and to supply raw materials for industrial production.

In view of weather and topographical conditions, and of the virtual nonexistence of new arable land, low yields,[22] low pace of technological change, and shortage of manpower, production figures as shown in Table 7.5 appear to be quite inflated, perhaps by as much as 25 to 30 percent. Between 1955 and 1963, when official statistics were available, yields varied from 2.3 tons per hectare (1959 bumper crop) to 1.6 tons (1955) with an average yield for the period amounting to around 1.9 tons. In the mid-1960s an ambitious goal of 5 tons per hectare (two crops) was called for, and it is claimed that by 1972, 297,000 hectares (45 percent of the two-crop land) under cultivation had already reached that goal,[23] while the number of cooperatives achieving the goal increased to 3,763 (22 percent of all cooperatives) as against 640 in 1965 (5 percent).[24]

As shown in Table 7.5, production of paddy (unprocessed rice) fluctuates widely from year to year. From 3.5 million tons in 1955, output reached 4.6 million tons in 1958, the first year of the three-year plan. With the exception of the 1959 bumber crop, when 5.2

TABLE 7.5

North Vietnam: Production of Food Grains, 1955-76
(in thousand metric tons)

Food Grain	1955	1956	1957	1958	1959	1960	1961	1962	1963	1964[a]	1965-68 (average)[a]	1969-71 (average)[a]	1972[a]	1973[a]	1974[b]	1975[b]	1976[b]
Rice (unprocessed)																	
Autumn crop	1,163	1,584	1,473	1,224	1,767	1,161	1,719	1,614	1,565	—	—	—	—	—	—	—	—
Spring and summer	—	—	—	—	—	110	134	131	120	—	—	—	—	—	—	—	—
Winter	2,360	2,552	2,475	3,353	3,426	2,941	2,798	2,797	2,611	—	—	—	—	—	—	—	—
Total	3,523	4,136	3,948	4,577	5,193	4,212	4,651	4,542	4,296	4,512	4,329	4,325	5,121	4,644	5,276[b]	4,867[c]	5,500[b]
Percentage change	—	17	-5	16	13	-19	10	-2	-5	5	-4	—	18	-9	—	—	—
Processed rice equivalent[d]	2,290	2,688	2,566	2,975	3,375	2,738	3,023	2,952	2,792	2,933	2,814	2,811	3,329	3,019	3,429	3,164	3,575
Subsidiary crops																	
Sweet potatoes	535	866	540	424	719	584	985	860	1,279	—	1,360	1,018	911	802	968	—	—
Maize	186	243	197	192	246	215	275	276	217	—	—	—	—	—	—	—	—
Manioc	163	367	186	154	260	335	729	883	864	—	1,037	906	1,039	919	—	—	—
Beans	11	18	21	21	22	11	17	17	13	—	—	—	—	—	—	—	—
Total, subsidiary	895	1,494	944	791	1,247	1,145	2,006	2,036	2,373	2,138	2,397	1,924	1,950	1,721	1,700[e]	1,564[e]	1,970[b]
Total, food grains	4,418	5,630	4,892	5,368	6,440	5,357	6,657	6,578	6,669	6,650	6,726	6,249	7,071	6,365	6,976	6,431	7,470

[a]Reconstructed North Vietnamese data based on growth rates. These averages may not agree with figures in Table 5.2 because of differences in methods of estimation and in sources of data; for 1964, estimates were based on area under cultivation (2,426,000 hectares) and average yield (1.9 tons per hectare).

[b]Planned level.

[c]Estimated as 13 percent less than the 1976 planned level.

[d]One kilogram of unprocessed rice (or paddy) gives an average of 0.650 kilograms of rice, after husking.

[e]Estimated based on growth rates implied in annual reports.

Sources: DRVN, Results of Agricultural Production During Fifteen Years Under the DRVN (Thanh Tich San Xuat Nong Nghiep Trong 15 Nam Duoi, Che Do VNDCCH) (Hanoi, 1960), pp. 58-59; DRVN, So Lieu Thong Ke (Statistical data), 1955-59 (Hanoi), p. 131, 1963, pp. 64-65; DRVN, Etudes Vietnamiennes, no. 2, 1964, p. 180, no. 13, 1967, pp. 6-21, no. 27, 1970, pp. 8-14, no. 44, 1976, pp. 171-75; Le Thanh Nghi, "Economic Situation in the North and the Tasks and Plan for 1976" (Report given at the Second Session of the National Assembly, Fifth Legislature, December 1975), Hoc Tap, no. 1, 1976, pp. 7-31; DRVN, Le Courrier du Viet Nam (Monthly bulletin), no. 45, February 1976, pp. 4-6; Vo Nhan Tri, Croissance Economique de la Republique Democratique du Vietnam (1945-65) (Hanoi: Editions En Langues Etrangeres, 1967), p. 423.

million tons were reportedly harvested, the 4. 6-million-ton mark appears to be the ceiling of North Vietnam's annual output throughout the 1960s. For the period of 1961-64, output averaged 4. 5 million tons per annum. According to the most recent data, average annual production during 1965-71 amounted to only 96 percent of the 1961-64 level. In 1972, output rose 18 percent from the 1969-71 average, then dropped 9 percent in 1973. On the basis of the 4. 5-million-ton average for 1961-64, these percentages may be reconstructed to imply output levels of 4. 3 million tons for 1965-71, 5. 1 million tons for 1972, and 4. 6 million tons for 1973. Paddy-output targets planned for 1974 and 1976 were 5. 3 million tons and 5. 5 million tons, respectively. Actual output for 1975 was reported to be 13 percent less than the planned 1976 level (4. 9 million tons).

As mentioned earlier, these figures have to be substantially discounted, and the 4. 2- to 4. 6-million-ton range for the 1960s can only be taken as indicative. Based upon internal consistency, crop yields, and land under cultivation, output in the 1960s, according to the author's own estimate, vary between 3 million and 3. 5 million tons per annum. This estimate appears to correspond also to the food-supply availability in North Vietnam.[25]

On the basis of a conversion ratio of one kilogram of paddy to 0. 65 kilograms of processed rice, the rice output averaged only 2. 9 million tons for 1960-64, 2. 8 million tons for 1965-71, and an estimated 3. 3 million tons for 1972-76.

With a fast-growing population (a rise of over 3 percent per annum), per-capita output appears to have declined steadily. Even taking the official data on paddy, estimated per-capita output shrank from 282 kilograms in 1961 to around 206 kilograms in 1968 and to 186 kilograms in 1973.[26] Bad weather in 1975, which destroyed 70, 000 hectares of the tenth-month crop alone, obviously reduced the per-capita availability to a record low.

Subsidiary Crops

Failing to increase the rice output sufficiently to meet the increasing requirements of the population, the government has encouraged the production and consumption of subsidiary crops. At the state's stores, when the rice stock is not adequate to meet demand, people who hold rice ration coupons and who want to buy rice are offered substitutes in maize, sweet potatoes, manioc and beans. The conversion rates from subsidiary crops to rice equivalence offered to customers have been fixed at 1 kilogram of rice for 0. 7 kilograms of maize, for 2. 3 kilograms of sweet potatoes, for 2. 3 kilograms of manioc, or 0. 5 kilograms of beans.[27]

TABLE 7.6

Production of Main Industrial Crops

(in tons)

Crop	1955	1957	1960	1961-64 (average)	1965-68 (average)[a]	1972[a]
Jute	1,121	5,700	12,440	10,785	14,869	11,556
Soybeans	5,018	7,002	14,885	8,968	10,044	5,201
Groundnuts	14,000	23,740	26,038	29,763	33,164	38,479
Sugar cane	100,000	348,353	451,701	543,033	507,030	477,029
Tobacco	600	1,730	1,655	4,781	6,667	8,645
Tea	2,300	2,637	2,650	3,035	6,803	9,686
Coffee	172	345	250	256	655	1,907
Cotton	2,160	5,721	4,623	—	4,000	—

[a]Reconstructed North Vietnamese data.

Sources: DRVN, Etudes Vietnamiennes, no. 13, 1967, p. 15, no. 38, 1974, p. 85, no. 44, 1976, p. 176.

Output of subsidiary crops increased substantially in the late 1950s (subsidiary crops are dry crops and are less vulnerable to floods and drought). It reached 2.4 million tons in 1963 as against 895,000 tons in 1955, with sweet potatoes and manioc accounting for most of the increase.

In 1963 output of sweet potatoes was already more than double the 1955 level, reaching 1.3 million tons. Production during 1965-68 remained roughly at the 1963 level. Thereafter, it followed the same declining trend as paddy output, as 1973 production amounted to only 77 percent of the 1961-64 average. Second in importance in terms of quantity is manioc, which amounted to 864,000 tons in 1964, a fivefold increase over the 1955 production.

Taking rice and subsidiary crops together, the production of food grains in North Vietnam may be estimated, according to reconstructed data, to have reached a record high of 7 million tons in 1972, representing only a 7 percent increase over the output level of a decade earlier. During the same decade, population grew by around 30 percent. The year 1972 is considered a very good year compared to 1971, 1973, and 1975, when production is estimated to be around 6.4 million tons.

Industrial Crops

Table 7.6 shows the production of main industrial crops. In order to supply raw material for industry and for export, industrial crops have received more and more attention in North Vietnam's development programs. In the 1960s, over 200,000 hectares, or 6 percent of total cultivated land, was reserved for industrial crops, mainly in the highland and the mountainous regions. In the late 1950s special effort was given to the expansion of cotton, and output increased to nearly 6,000 tons in 1957, or by 165 percent in only two years. Cotton output appeared to level off during the 1960s, and average output for 1965-68 was around 4,000 tons. During the same period other crops, especially coffee, tea, and tobacco, increased steadily. In 1972, tea and coffee combined amounted to 12,000 tons, and tobacco, nearly 9,000 tons, a fourfold and fivefold increase over the 1960 level, respectively.

During 1965-68 industrial crops accounted for nearly 7 percent of total value of agricultural output as compared to 4 percent during 1955-60.[28] The share could have been higher had it not been for the reduction of area under industrial-crop cultivation in recent years in favor of food crops. It is reported that whenever the rice crop is bad, the peasants subsequently save the land to grow rice and other food crops even in areas that are reserved for industrial

TABLE 7.7

Number of Animals, 1939-74
(thousand heads)

Year	Animal			
	Buffaloes	Oxen	Pigs	Poultry
1939	788	563	2,255	—
1955	1,084	756	2,137	—
1957	1,238	906	2,950	—
1959	1,447	950	3,626	41,759
1960	1,451	879	3,751	47,469
1961	1,450	871	3,740	42,579
1962	1,480	786	4,239	43,611
1963	1,559	796	4,491	57,683
1969[a]	1,655	720	5,486	71,204
1972[a]	—	—	5,579	—
1973[a]	—	—	6,140	—
1974[a]	1,395	997	6,382	—

[a]Reconstructed North Vietnamese data.

Sources: DRVN, Etudes Vietnamienne, no. 2, 1964, pp. 180-81, no. 13, 1967, pp. 16-17, no. 44, 1976, pp. 175-77; DRVN, La Republique Democratique du Viet-Nam (Hanoi: Editions En Langues Etrangere, 1975), pp. 88-89.

crops; in these areas the farmers are not supposed to grow rice because they have already received food subsidies from the state.[29]

The 1976 plan set a target to increase area under industrial-crop cultivation by 20 percent over the 1975 level, especially the area devoted to sugar, malt, jute, hemp, and peanuts.[30]

Animal Husbandry and Fishing

Animal husbandry and fishing play a crucial role in the government's agricultural-diversification policy. The Fifth Plenum of the party's Central Committee held in 1961 laid special emphasis on cattle raising to increase meat consumption of the population and to supply organic fertilizer for agriculture.[31] Perhaps, farming and fishing are the only activities in agriculture where the government can still tolerate a certain amount of private ownership, since animal husbandry is widely undertaken by both cooperatives and individual households. The government allows each peasant family to retain 5 percent of its cultivated land (by transferring to cooperatives) to raise domestic animals, especially pigs, and for the growing of fruits and vegetables for family consumption.[32] Above the private lot, each cooperative is also allowed to retain 5 percent of its land and 50 percent of sweet-potatoes and manioc output for cattle raising. It is estimated that an average cooperative having 200 hectares of land would be able to raise between 250 and 300 pigs on 10 hectares of land. Highest priority is attached to the raising of pigs as they are the major source of meat for the population and an important source of fertilizer. In addition to the provision of 5 percent landholding, pig raising also receives other incentives: the farmers are paid fair prices even when sell to the state,[33] and they are allowed to keep between 50 and 60 percent of pork for private consumption or to sell on the free market (in case they signed a contract to sell meat to the state).[34] As a result of incentives, the number of pigs increased from 2.1 million in 1955 to an estimated total of 5.5 million in 1969 and 6.4 million (more than triple) in 1974—representing perhaps the most successful activities in North Vietnam's agricultural sector (see Table 7.7).

Next to pig raising, water buffaloes, a traditional source of draft power, also increased steadily but at a much slower pace, reaching an estimated 1.7 million heads in 1969 as compared to 1.1 million heads in 1955. The change in cattle heads was relatively insignificant during the period, with a declining trend beginning in 1960; the 1969 cattle population was 76 percent of that in 1959.

Fishing is another source of private income. Thanks to a large network of rivers all over the country and to the paddy fields, inland fishing provides an excellent opportunity for peasants to supplement their food rations and to earn side income. A growing number of farm cooperatives have also been set up to undertake large-scale fish raising. Area reserved for fish raising has increased from 17,000 hectares in 1959 to 63,000 in 1960. In the following year it was almost doubled to 122,000 hectares. In 1967 the value of fish output was 4 percent of that of agricultural production as compared to 1.4 percent in 1958. In the same year, the per-capita supply of fish was 6 kilograms with some cooperatives having as high a supply as 20 and 50 kilograms.[35] Fishing activities also earned up to 36 percent of cooperatives' income in some provinces.[36]

CONCLUSION

Analysis of the agricultural sector in this chapter shows that the Northern food problem, instead of being alleviated, has been exacerbated since 1954. Paddy production, which grew at a modest rate of 3.2 percent a year on average during 1955-64, as compared to 3 to 3.5 percent annual population growth, was characterized by sharp fluctuations with rates as high as 14 percent (1959) and as low as 19 percent negative (1960). For the rest of the 1960s and the early 1970s, no significant progress was visible.

The agricultural problem appears even more dramatic when paddy output is converted to processed rice ready for consumption. Rice output reached its peak in 1959 when 3.4 million tons were produced. With the exception of three years (1961, 1972, and 1973), actual production never exceeded the 3-million-ton peak. The rice consumption, though carefully rationed, has had to be supplemented by other food crops such as sweet potatoes and manioc and by imports estimated at 1 million tons a year.

Leaving the war aside, the failure in agriculture is explained partly by natural conditions, partly by the lack of incentive due to collectivization, and partly by the inadequate investment in agriculture resulting from emphasis on heavy industry as analyzed in the next chapter.

NOTES

1. In the winter, under the cloudless sky, which permits higher intensity of sunlight transmission and absorption, conditions for increasing rice yields are more favorable than in the summer season when the sky is overcast with tropical monsoon clouds. See "Planning Strategies in Agriculture," Economic Bulletin for Asia and the Far East (United Nations) 20, no. 2, (September, 1969) pp. 12-13.

2. Le Thanh Nghi, "Nhiem Vu, Phuong Huong Khoi Phuc Kinh Te Hai Nam 1974-75 Va Ke Hoach Nha Nuoc 1974" (Tasks and orientation of economic reconstruction during 1974-75 and state plan for 1974), Nghien Cuu Kinh Te (Economic studies), no. 77, February, 1974, p. 3.

3. DRVN, So Lieu Thong Ke (Statistical data) 1955-59, (Hanoi, 1960), p. 57.

4. According to North Vietnam's latest regional data, a slightly different picture is given (indicating percent of the population

and 74 percent of cultivable land). See Le Duan and Pham Van Dong, Vers Une Agriculture de Grande Production Socialiste (Hanoi: Editions En Langues Etrangeres, 1975), p. 96.

5. Vo Nhan Tri, Croissance Economique de la Republique Democratique du Vietnam (Hanoi: Editions En Langues Etrangeres, 1967), p. 479.

6. A detailed picture on agricultural zoning for 1985 is given by Bui Huy Dap in "Nhung Bien Doi Cach Mang Sau Sac O Nong Thon va Trong Nong Nghiep Mien Bac" (Revolutionary and profound changes in North Vietnam's rural area and in agriculture), Khoa Hoc va Ky Thuat Nong Nghiep, no. 2, 1970, pp. 77-78.

7. For a good description of the spring crop, see Duong Hong Hien, "La Structuration des Cultures," in Etudes Vietnamiennes, no. 38, 1973, pp. 28-30.

8. Ibid., p. 39.

9. DRVN, Etudes Vietnamiennes, no. 44, 1976, p. 154.

10. Ibid., p. 37.

11. DRVN, Etudes Vietnamiennes, no. 38, 1973, p. 43.

12. Empirical studies on irrigation in the context of Southeast Asian agriculture (which is more relevant for North Vietnam) suggest that, in countries where land productivity is still low (less than 2.5 tons per hectare), irrigation is indeed a leading input. The role of irrigation in successive productivity stages is, first, to stabilize the harvest fluctuations due to deficient or untimely rainfall; second, to make possible the introduction of a second crop; and third, to make possible the increased application of fertilizer, better seeds, and improved farming techniques. See Shigeru Ishikawa, Economic Development in Asian Perspective (Tokyo: Institute of Economic Research, Hitotsubashi University, 1967), pp. 84-122; K. Takase and T. Kano, "Development Strategy of Irrigation and Drainage," Asian Agricultural Survey (Asian Development Bank, Manila) 2, 1968, pp. 600-36.

13. Between 50 and 60 percent of fertilizer used in the cooperatives comes from human and animal excrement. Persons over 15 years of age are required to collect 110 pounds of excrement and 1,320 pounds of urine each year. The annual quota for water buffaloes is 3.5 tons of fermented manure, for cattle, 3 tons, and for pigs, 2.5 tons. See Jon M. Van Dyke, North Vietnam's Strategy for Survival, (Palo Alto: Pacific Books, 1972), p. 187.

14. DRVN, Etudes Vietnamiennes, no. 38, 1973, p. 126.

15. Ibid., p. 135.

16. DRVN, Etudes Vietnamiennes, no. 44, 1976, p. 157.

17. The policy of agricultural mechanization was first laid down by the party's Third National Congress as follows (DRVN, Etudes Vietnamiennes (English ed.), October 1965, p. 93):

> State farms must achieve mechanization, with modern machines; co-operatives must strive to increase the numbers of draught animals as much as possible, they must forge ahead with the improvement of farm implements and transport means, widely put into practice the method of ploughing with two buffaloes or oxen, actively use different sorts of semi-mechanized farm implements, through which mechanization may gradually be attained. In the Five-Year Plan period a number of agricultural machinery stations must be set up at given places with the aim of spreading the experimental use of agricultural machines, and of partly solving our difficulties in the shortage of tractive power.

On the supplement of females to the male labor force during war time and the role of mechanization, see Bui Huy Dap, "Small Agricultural Mechanization in the D. R. V. N.," DRVN, Etudes Vietnamiennes (English ed.), no. 13, 1967, pp. 68-69; Hoc Tap September 1967, pp. 48-55, 89; Le Monde, November 25, 1967, p. 3.

18. DRVN, Etudes Vietnamiennes, no. 44, 1976, p. 161.

19. Productivity of a mechanical thresher is three to five times higher than that of hand- or buffalo-driven threshers and the cost is 20 to 60 percent lower; output of a hydraulic pump with a 273-cubic-meter-per-hour capacity is equivalent to that of 80 persons. See Bui Huy Dap, op. cit., pp. 70-71.

20. DRVN, Etudes Vietnamiennes, no. 44, 1976, p. 160.

21. Pham Van Dong, Realisations et Taches Immediates (Hanoi: Editions En Langues Etrangeres, 1955), pp. 64-65; DRVN, Plan Triennal (Hanoi: Editions en Langues Etrangeres, 1959), p. 36.

22. Average yields prior to partition were very low: 1.3 tons per hectare for 1939 (a good year), 1.2 tons (1942) and 1.4 tons (1954). Thus, it is difficult to believe that yield could have reached 2 tons per hectare only two years after partition. For data on yield by province in a typical prepartition year, see Table 1.4. For

information on yields since partition, see Vo Nhan Tri, op. cit., pp. 215, 291, 423.

23. On the goal of five tons per hectare, see Cuong Pham, "Five Tons Per Hectare?," in DRVN, Etudes Vietnamiennes, October 1965, pp. 82-99. Compared to 1965, the 1972 result (297,000 hectares having a yield of 5 tons each) was a ten-fold increase (DRVN, Etudes Vietnamiennes, no. 44, 1976, p. 174).

24. Ibid.

25. For example, the per-capita food supply ration was 13.5 kilograms per month, or 162 kilograms per year in 1966 (Harvey H. Smith, et al., Area Handbook for North Vietnam [Washington, D.C.: U.S. Government Printing Office, 1967], p. 326); On the basis of population in that year (19.5 million), total food-grain availability would be 3.2 million tons including rice and other food grains. Assuming that the share of rice in total food-grain consumption was 70 percent, total rice availability in 1966 was 2.2 million tons, or equivalent to 3.4 million tons of paddy. This level of paddy availability must correspond roughly to actual output in the same year, which in turn should have been the criterion upon which the per-capita ration was fixed.

26. On the basis of standard minimum requirement for an average Vietnamese, which is about 240 kilograms per year, the 1961 per-capita output indicates a per-capita surplus of 42 kilograms as compared to a deficit of 34 kilograms in 1968 and 54 kilograms in 1973. The latter figure implies a total deficit of 1.3 million tons in that year (assuming a population of 24 million). About the rice requirement, see Robert L. Samson, The Economics of Insurgency (Cambridge: MIT Press, 1970), pp. 37-38.

27. DRVN, Etudes Vietnamiennes, no. 13, 1967, p. 8.

28. DRVN, Etudes Vietnamiennes, no. 44, 1976, p. 176.

29. DRVN, Etudes Vietnamiennes, no. 38, 1973, p. 83.

30. Le Thanh Nghi, "Economic Situation in the North and the Tasks and Plan for 1976" (Report given at the Second Session of the National Assembly, Fifth Legislature, December 1975), Hoc Tap, no. 1, 1976, pp. 17-18.

31. Nguyen Lao Dong, "L'Elevage," in DRVN, Etudes Vietnamiennes, no. 27, 1970, p. 50.

32. For a family of five persons, the lot is about 300 square meters. (ibid., p. 53).

33. Ibid., p. 61.

34. Ibid., p. 59.

35. It is estimated that in the future, the per-capita supply of fish may be increased by fourfold (ibid., p. 79).

36. This compares with only 15 percent five years earlier (1962) (ibid., p. 68).

CHAPTER

8

THE INDUSTRIAL SECTOR

Following the experience of the Soviet Union and of China in building socialism, North Vietnam's leaders have never ceased to insist on rapid industrialization with top priority to heavy industry.[1] Nevertheless, systematic industrialization only began in 1960 with the preparation of the first five-year plan.

During the first three years of the regime (1955-57) priority was given to the restoration of infrastructure and agriculture. Industrial development during the three-year plan (1958-60) focused on both agriculture and industry with attention given to the production of basic consumer goods (food, clothing, and education).[2] Clear direction toward rapid development of heavy industry was given by the Third Congress of the Worker's party in September 1960, which laid down the task and orientation of the first five-year plan as follows:

> To build a balanced and modern socialist economic structure, to develop both industry and agriculture and take heavy industry as the basis, to give priority to rational development of heavy industry and at the same time to strive to develop agriculture and light industry, to turn our backward agricultural country into a country endowed with a modern industry and agriculture.[3]

With substantial foreign resources and an important number of technicians from the Soviet Union, China, and other socialist countries, industry expanded at a rapid pace during 1961-64 with the building of several new and important factories. Industrialization was slowed down considerably during the air war (1965-68), recovered somewhat during 1969-71, but declined again during 1972. It is claimed that by the end of 1975, the industrial base had been nearly restored to the 1965 level.[4]

The rationale underlying North Vietnam's choice to concentrate almost entirely on the development of industry, as well as the role of industry in the overall plans, were discussed in Chapter 5. This chapter examines the execution of that policy and the state of industrial development.

STRUCTURE OF INDUSTRY

Industry is divided into three main categories: groups (A and B), sectors (industry and handicrafts), and administrations (central and regional).

In the first five-year plan (1961-65), the classification of industrial production into group A and group B was significant because of the role of heavy industry in the plan. During the war years of 1966-73, the classification into central and regional administrations was more meaningful as the centrally located factories were either destroyed or were dispersed into the countryside. In this period, the regional industry became increasingly important in terms of both production and state investment.

Group A, also called heavy industry or production-goods industry, includes electricity, engineering, metallurgy, chemicals, mining, and construction materials. Group B, or light industry, includes consumer goods such as food products, textile and textile products, paper and paper products, and wood processing.

During peacetime conditions, centrally run industry is conceived to satisfy both short-term and long-term requirements of the national economy in electricity, machinery, mining, and metal works, while regional industry is confined to the supplying of goods for local consumption such as tools and other inputs for local agriculture (fertilizer, insecticides) and for export.[5] To a large extent central industry is the same as heavy industry (group A), plus some light industry in group B, while regional industry comprises mainly handicrafts and an important part of light industry (group B).

Official data on the gross value of industry and handicraft production (1955-63) and data reconstructed by the author (1964-75) are presented in Table 8.1. Again, these figures may only be taken as showing some magnitude rather than as showing growth in absolute terms, due to the fact that they may not always be computed on a value-added basis and, therefore, double counting most certainly was involved.

TABLE 8.1

Value of Industrial and Handicraft Production, 1955–75
(in millions of dong, constant 1959 prices)

	Group			Sector			Administration		
Year	A (heavy industry)	B (light industry)	Total	Industry	Handicrafts	Total	Central	Regional	Total
1955									
Value	86	231	317	92	225	317	38	279	317
Percent	27	73	100	29	71	100	12	88	100
1956									
Value	180	447	627	257	370	627	157	470	627
Percent	29	71	100	41	59	100	25	75	100
1957									
Value	190	644	834	345	489	834	246	588	834
Percent	24	76	100	43	57	100	28	72	100
1958									
Value	258	755	1,013	486	527	1,013	372	641	1,013
Percent	25	75	100	48	52	100	36	64	100
1959									
Value	425	886	1,311	708	603	1,311	623	688	1,311
Percent	32	68	100	54	46	100	48	52	100
1960									
Value	503	968	1,471	838	633	1,471	581	890	1,471
Percent	34	66	100	57	43	100	40	60	100
1962									
Value	788	1,313	2,101	1,366	735	2,101	977	1,094	2,101
Percent	37	63	100	65	35	100	47	53	100
1963									
Value	893	1,367	2,260	1,514	746	2,260	1,084	1,153	2,260
Percent	39	61	100	67	33	100	48	51	100
1964									
Value	1,044	1,567	2,611	1,828	783	2,611	1,354	1,257	2,611
Percent	40	60	100	70	30	100	52	48	100
1965									
Value	1,217	1,634	2,851	2,097	754	2,851	1,534	1,317	2,851
Percent	43	57	100	74	26	100	54	46	100
1966									
Value	1,278	1,493	2,771	1,985	786	2,771	1,418	1,353	2,771
Percent	46	54	100	72	28	100	51	49	100
1967									
Value	1,053	1,352	2,405	1,671	734	2,405	1,086	1,319	2,405
Percent	44	56	100	69	31	100	45	55	100
1968									
Value	1,149	1,399	2,548	1,852	696	2,548	1,193	1,355	2,548
Percent	45	55	100	73	27	100	47	53	100
1971									
Value	1,445	2,037	3,482	2,652	830	3,482	1,707	1,775	3,482
Percent	41	59	100	76	24	100	49	51	100
1972									
Value	1,245	1,841	3,086	2,272	814	3,086	1,398	1,688	3,086
Percent	40	60	100	74	26	100	45	55	100
1973									
Value	1,539	2,134	3,673	2,694	979	3,673	1,619	2,054	3,673
Percent	42	58	100	73	27	100	44	56	100
1975									
Value	—	—	4,940	—	—	—	2,481	2,459	4,940
Percent	—	—	100	—	—	—	51	49	100

Note: Individual items may not add up to sum because of rounding.

Sources: For 1955–63: DRVN, So Lieu Thong Ke, 1955–59, (Hanoi, 1960) pp. 154–58, 160, So Lieu Thong Ke, 1960–63 (Hanoi, 1963) pp. 43–45; Vo Nhan Tri, Croissance Economique de la Republique Democratique du Vietnam (1945–65) (Editions En Langues Etrangeres, 1967), p. 229. For 1964–75: DRVN, La Republique Democratique du Vietnam (Hanoi: Editions En Langues Etrangeres, 1975), p. 79; Leon Lavallee, L'Economie du Nord Vietnam (Paris: Centre d'Etude et de Recherche Marxiste, 1971), p. 50; DRVN, Etudes Vietnamiennes, no. 44, 1976, pp. 200, 253.

During the first ten years of development (1955-65), especially during 1961-65, North Vietnam's industry underwent a marked structural change reflecting the government's all-out campaign to industrialize the backward and predominantly agricultural economy. The share of industry in the total value of industry and handicrafts output reached 70 percent in 1965 as compared with 29 percent ten years earlier. The substantial increase was due to the fast growth of industry under group A, especially electric power, mining, tool manufacturing, and chemical production. The relative share of group A in the gross value of industrial output rose to 43 percent in 1965 as against 34 percent in 1960 and 27 percent in 1955. It is obvious that the rise took place at the expense of group B (consumer goods), which declined steadily from 76 percent in 1957 to 54 percent in 1966. The growth of group A also implies the expansion of centrally run industry.

With the elimination of private enterprises, all industrial establishments in North Vietnam are owned and controlled by either the centrally run state enterprises or by the regional industrial cooperatives (see Chapter 4). The share of centrally run industry, which accounted for only 12 percent of total industrial production value in 1955, rose quickly to 48 percent in 1959. At the end of the first five-year plan, it exceeded the share of regional output by 5 percent.

With the air war beginning in 1966, the structural change of industry took a sharp turn and was reversed to the late 1950s and early 1960s pattern. Although the government insisted that it did not abandon the goal of industrialization—and in fact the output value of group A during 1966-71 was slightly higher than the 1965 level—large-scale, centrally run industries were definitely receding, rendering the handicraft, regional sector increasingly important. The share of handicrafts in total industrial and handicraft output value rose to 31 percent in 1967 as compared to 26 percent in 1965. It declined to 21 percent in 1971 as a result of two years' reconstruction during which modern industry recovered somewhat.

The declining role of centrally run industry since 1966 is partly because of the impact of bombing and partly because of the government's policy of dispersion of industry to the provinces. The share of regional industry in total industrial output value increased from 46 percent to 55 percent in only two years (1966-67) as the value of central industry's output declined by 30 percent.[6]

Regional industry continues to play an increasingly important role, and by 1973, its share in total industrial output value reached 56 percent as against 51 percent in 1971. The change no doubt reflects the effects of the 1972 air war. Estimates for 1975 show equal shares for both regional and central industry.

TABLE 8.2

Budgetary Investment in Industry, 1955–71
(annual average in millions of dong, constant 1959 prices)

Industry Group/	1955–57		1958–60		1961–64		1965–68		1969–71	
Administration	Amount	Percent	Amount	Percent	Amount	Percent	Amount	Percent	Amount	Percent
Group										
A	44	66	142	67	308	78	304	82	350	76
B	23	34	70	33	87	22	67	18	111	24
Total	67	100	212	100	395	100	371	100	461	100
Administration										
Central	67	100	208	98	371	94	278	75	346	75
Regional	—	—	4	2	24	6	93	25	115	25
Total	67	100	212	100	395	100	371	100	461	100
Investment in industry as percent of total development investment										
Group A		20		31		41		27		30
Group B		11		15		12		6		10
Total		31		46		53		33		40

Sources: DRVN So Lieu Thong Ke, 1955-59 (Hanoi, 1960), p. 99, So Lieu Thong Ke, 1967 (Hanoi, 1967), p. 13; DRVN, La Republique Democratique du Vietnam (Hanoi: Editions En Langues Etrangeres, 1975), pp. 67, 76–79; Etudes Vietnamiennes, no. 44, 1976, p. 196; Leon Lavallee, L'Economie du Nord Vietnam (Paris: Centre d'Etude et de Recherche Marxiste, 1971), p. 58.

STATE INVESTMENT IN INDUSTRY

The growth and distribution of budgetary investment in industry is shown in Table 8.2. In absolute terms, investment in industry rose rapidly during the first ten years of development (1955-65), reflecting the planners' priority in industrialization. From an annual average of 67 million dong during 1955-57, industrial investment increased to 212 million dong during the first three-year plan and 395 million dong during the five-year plan. In terms of resource allocation, these investments accounted for 46 percent and 53 percent, respectively, of total state investment in the economic sector. Beginning with the air war in 1965, industrial share of budgetary investment declined sharply to 33 percent (average for 1965-68) and to 40 percent (average for 1969-71), as a substantial portion of public funds was diverted to the repair and construction of the transport and communication network (52 and 60 percent, respectively).[7]

Among sectors, investment in heavy industry (group A) occupied an overwhelming portion of state industrial investment at the expense of consumer-goods industries, averaging 66 percent during the second half of the 1950s and 79 percent during the 1960s.

Reflecting the planners' view that local industry is to be financed by local investment in terms of manpower and financial resources, nearly the total of the state-investment fund went to centrally run projects during the first ten years of development. Since 1965, however, with the emphasis on regional industry, a substantial portion of state resources (25 percent) was allocated to finance regional production.

TREND IN INDUSTRIAL DEVELOPMENT

Table 8.3 shows the general trend of production of major industrial commodities that make up the total industrial output. Generally speaking, the output level of major products increased rapidly during 1955-65, especially during the first five-year plan. All industry suffered severe decline during the 1965-68 bombing campaign, recovered somewhat during 1969-71, and then experienced another setback during 1972 as a result of total mobilization of resources for the offensive in the South and because of U.S. bombing in the North. It is claimed, however, that the 1973-75 reconstruction program successfully restored the industrial base to the 1965 level.[8]

TABLE 8. 3

Output of Major Industrial and Handicraft Products, 1955-75

Product	1955	1956	1957	1958	1959	1960	1962	1963	1964	1965	1968	1971	1972	1973	1975
Electric power (million kilowatt hours)	53	94	121	160	209	256	368	460	595	634	376	759	550	804	1,320
Coal (thousand tons)	642	1,214	1,085	1,549	2,202	2,595	3,500	3,300	3,600	4,201	2,401	3,402	1,699	2,297	4,644
Apatite (thousand tons)	—	24	66	136	261	490	678	930	846	680	—	218	195	173	—
Cement (thousand tons)	9	197	165	302	381	408	462	491	600	573	70	353	155	229	390
Chromite (thousand tons)	—	1	4	6	7	19	33	31	29	32	—	—	—	—	—
Water pumps (each)	—	—	—	—	—	—	1,367	1,966	2,064	2,524	2,138	918	768	685	—
Chemical fertilizer (thousand tons)	6	34	23	32	51	57	108	296	159	144	119	235	179	194	382
Timber (thousand cubic meters)	362	457	439	459	770	753	941	992	1,107	1,085	722	866	752	693	—
Paper (thousand tons)	1	2	2	3	4	5	15	17	20	24	9	19	11	12	—
Cotton cloth (million meters)	9	48	63	66	78	76	95	80	107	100	83	109	73	76	88
Silk (million meters)	—	2	4	7	6	6	7	10	13	—	—	—	—	—	—
Sugar (thousand tons)	2	5	4	6	11	10	19	25	27	31	14	14	13	14	—
Salt (thousand tons)	95	88	110	125	130	117	144	127	151	153	173	198	150	164	—
Cigarettes (million pack)	3	12	15	30	66	73	95	105	109	134	134	175	164	220	—
Soap (thousand tons)	1	2	3	3	4	5	6	6	7	8	4	4	5	6	—
Bicycles (thousand tons)	—	—	—	6	13	27	39	60	78	85	86	62	61	61	—

Sources: For 1955-63: DRVN, So Lieu Thong Ke, 1955-59 (Hanoi, 1960), pp. 163-64, So Lieu Thong Ke, 1960-63 (Hanoi, 1963), pp. 50-51. For 1963-75: DRVN, Etudes Vietnamiennes, no. 44, 1976, pp. 200-02; Vo Nhan Tri, Croissance Economique de la Republique Democratique du Vietnam (1945-65) (Hanoi: Editions En Langues Etrangeres, 1967), p. 320; Leon Lavallee, L'Economie du Nord Vietnam (Paris: Centre d'Etude et de Recherche Marxiste, 1971), p. 50. Le Monde, December 25, 1970; Far Eastern Economic Review, 1972 Yearbook, pp. 255-59.

Heavy Industry

Electricity and Fuel

As a basic input of mechanized industry, electricity received the highest priority on North Vietnam's list of industrial projects.[9] An ample supply of coal helped to expand the electric-power sector very rapidly—representing perhaps the most impressive industrial undertaking in the entire history of the country's economic development. Electric-power generation increased from 53 million kilowatt hours in 1955 to 256 million in 1960 and to 634 million estimated for 1965. It then drastically dropped to 376 million kilowatt hours estimated for 1968, or by 41 percent as compared to 1965. The 1969-71 reconstruction program vigorously restored the power industry and it is claimed that output in 1972 was 28 percent higher than output in 1964, implying a production of 759 million kilowatt hours. The 1973 level is said to have been 35 percent above the 1964 level, implying an output of 804 million kilowatt hours. An estimate based on the official growth index shows a 64 percent increase over a two-year period (1973-75), output reaching 1.3 billion kilowatt hours in 1975.

Most of the power plants are thermal plants. Among the most important ones are the Uong Bi (Quang Ninh province), the Thanh Hoa, Vinh, and Ninh Binh plants. The Uong Bi plant, built during the first five-year plan, is potentially the most important thermal plant because of its proximity to the Hon Gai coal bed. Although severely damaged during the air-war years, the plant was restored and expanded rapidly during the reconstruction years (1969-71 and 1973-75), and by 1973 a third extension was built producing an additional 100,000 kilowatt hours.[10]

There are a number of small hydroelectric-power stations throughout the highlands, but their output is still relatively small. An important project, the Thac Ba hydroelectric plant (Yen Bai province), was built during the first five-year plan to use the water flow of the Chay River and to generate an annual output of 500 million kilowatt hours by 1966, or twice as high as the country's 1964 total electric-power output.[11] The project was damaged and suspended during the 1965-68 war years, but is currently being reconstructed, and it is not clear when the plant will begin its full operation.

At the current rate of distribution, over 80 percent of electric power is allocated for industrial production, irrigation, and flood control, and less than 20 percent for consumption.[12]

Mining

Traditionally the mining center of all Vietnam, the Northern part of the country has been known throughout Southeast Asia for its richness in deposits of coal, apatite, iron ore, chromite, copper, tin, lead, and zinc. With the exception of coal, chromite, tin, and apatite, most of these minerals have not been fully exploited.

Coal, the most important mineral and export product in North Vietnam, is mined mainly at the Hong Gai basin (northeast of Hai Phong), which includes the Cam Pha, Quan Trieu, Lang Cam, and Long Quang mine fields. The basin stretches from the Ha Long Bay almost to Thai Nguyen province. During the colonial era, coal output (2.6 million tons in 1939, a peak year) was sufficient to meet domestic requirements and to be exported in large quantities. Coal mining was given high priority during the years of economic restoration, and 1956 output amounted to 1.2 million tons, or double the 1964 level. It increased steadily in the following years, reaching an annual average of 3.4 million tons during 1960-64, and an estimated peak of 4.2 million tons in 1965. Production declined sharply after 1965, and 1972 output was only 47 percent of the 1964 level, or about 1.7 million tons. The 1973 output rose appreciably to about 2.3 million tons, equivalent to the 1959 level, and an estimate for 1975, again based on the growth index, indicates an output doubling the 1973 level. During the first five-year plan, it was reported that about 1 million tons of coal, nearly one-third of the average annual output, was exported, mainly to Japan.[13]

An increasingly important mineral, apatite, a phosphate consisting chiefly of lime, has been produced since 1956. Apatite is being mined mainly in the Lao Cai province, the northern frontier bordering China, but the prospect for apatite also exists in Thanh Hoa province. The Cam Duong deposits in Lao Cai have an estimated reserve of one billion tons of phosphate and supply excellent input for the Lam Thao fertilizer plant.

Apatite output increased dramatically during the 1960-64 period as a result of the country's dire need for fertilizer. The peak was reached in 1963 when 930,000 tons were produced. The production trend was reversed in 1964, and the decline continued through the late 1960s and early 1970s. By 1973, only 173,000 tons were produced, a level approximating only 20 percent of the 1964 output.

High-grade chromite found in the mountainous area of Thanh Hoa province was mined in the early 1960s, producing an annual average of 31,000 tons. However, production was discontinued after 1965.

The Tri Cau iron ore deposits near Thai Nguyen, the most important mine in operation so far, has been exploited to supply ore to the Thai Nguyen steel mill.

Chemicals

Among the chemical industry's products the most important are fertilizer, insecticide, cotton, and silk. The first and biggest chemical-fertilizer plant is the Lam Thao superphosphate plant in Viet Tri province, using as input phosphate from the Cam Duong apatite mine. Besides the Lam Thao, there are about five other small plants. Total chemical-fertilizer production rose sharply in the first three years of the five-year plan, reaching a peak of 296,000 tons in 1963, then declined steadily to around 119,000 tons in 1968, or by 40 percent. After a rebound due to reconstruction during 1969-71, fertilizer output declined again after 1971. The 1975 output is estimated at 382,000 tons as compared to 235,000 tons in 1971, an increase of 63 percent. As already noted in the previous chapter, the production of other sources of fertilizer—the green fertilizer, an azolla type—has received great attention in recent years and has become an increasingly important component of total fertilizer output.[14]

Production of insecticides is still at the early stage of development, with a factory at Viet Tri adjacent to the Lam Thao phosphate plant. Its annual output is not yet known.

Metallurgy

Iron and steel lie at the heart of heavy industry and thus receive top priority in North Vietnam's long-run strategy of development. The construction of the Thai Nguyen iron and steel factory was considered the most important achievement of the five-year plan.[15] The factory is also the largest modern industrial enterprise in North Vietnam. It was designed to comprise three blast furnaces, a steel-rolling mill, an iron-smelting plant, and several other small shops complementary to steel and iron works. The first blast furnace was completed in 1963 and involved 20,000 workers and 38 months of work.[16] With an ample supply of iron ore nearby, the original annual output was planned to reach 200,000 tons of iron ingots. The plant was in operation only one year when it produced around 100,000 tons of pig and cast iron and steel. The factory was damaged during the bombing campaign of 1965-68, but was reconstructed thereafter, and by 1974 the third blast furnace was completely restored, with production resumed in 1975.[17]

Mechanical Works

During the first five-year plan, North Vietnam's mechanized industry moved from repair work to manufacturing.[18] A number of plants were set up to produce a variety of products: spare parts, tools, electric motors, simple agricultural equipment. The three largest factories are the Hanoi machinery factory, the Tran Hung Dao factory, also in Hanoi, and the Duyen Hai machine works in Hai Phong. The main products of these factories are small water pumps, diesel motors, electric motors, and transport equipment (small boats, towing equipment). The number of water pumps produced was very modest, estimated at 2,524 units in 1965, but an 84 percent increase over 1962, when they were first fabricated.

All areas of machinery production declined sharply during the 1965-68 period with a notable exception: the production of small agricultural machines, which increased 53 percent over the period.

Construction Materials

Cement, lime, bricks, tiles, concrete, and glass are the most important building materials produced. During the colonial era, the Hai Phong cement factory supplied cement to all of Indochina. Its peak output was reached in 1939 with 306,000 tons. The plant was enlarged in the early 1960s, and by 1964 produced 600,000 tons, or double the prewar level. Its output declined dramatically during the 1965-68 period, and by 1968 only 70,000 tons—12 percent of the 1964 level—was produced. In 1973, production was 38 percent of the 1964 level (229,000 tons).

Light Industry

Food Processing

Light industry includes food processing, textiles, and other consumer-goods production (cigarettes, rubber goods, soap, bicycles). Food-processing activities comprise mainly rice milling, fish and fish-sauce processing, and sugar refining. Rice milling represents a major technological improvement over the traditional manual husking method. Several rice mills were built throughout the country during the three-year and five-year plans. It is estimated that by 1963 over 339,000 tons of rice (12 percent of production) were processed by the mills.[19] Sugar refineries are situated at Van Dien, Viet Tri, and Vinh. Production capacity is estimated at 10,000 tons a year for the Van Dien plant and at 5,000 tons each for the Viet Tri and Vinh plants. Total sugar output reached 31,000 tons in 1965, which is triple the 1960 level. Since 1965, average output is about 14,000 tons.

Textiles and Other Consumer Goods

Textile manufacture is the largest and most important consumer-goods industry. Nam Dinh was the country's traditional center of textile production during the colonial era. In 1965 a new and modern plant was built in Hanoi, the March 8 textile mill. The plant was expected to produce at a capacity of 30 million meters of cloth annually, and consists of cotton-weaving, printing, and dying workshops. The Nam Dinh mill is the largest and has employed 12,000 workers and produced an output of 50 million meters of cloth in peacetime.[20] Total cotton-cloth output reached a peak in 1964 when 107,000 tons were produced. The 1973 output is estimated at 76,000 tons, or 71 percent of the 1964 level.

Bicycles, tires, soap, cigarettes, matches, and sugar are among the most important consumer products in addition to textiles. In 1964, 78,000 bicycles, 7,000 tons of soap, 109 million packs of cigarettes and 27,000 tons of sugar were produced, according to official statistics. With the notable exception of cigarette manufacturing, which continued to increase, most consumer-goods production declined sharply during 1965-68.

CONCLUSION

Analysis of industrial development in this chapter shows that, contrary to the agricultural situation, North Vietnam scored some impressive success in structurally changing the industrial base during the period of 1958 to 1965, especially during the first five-year plan. The growing role of industry can be seen in the share of value of industrial output in national income (in Table 6.5) and in the employment picture (see Table 8.4). However, the goal of rapid industrialization had to be abandoned after 1965 under the impact of the war and of the regime's decision to change national priority. The newly built industrial base had to be dismantled, the major factories were dispersed into the provinces. The centrally commanded economy had to give way to the local autonomous economy, and the heavy industry gave way to the light industry and handicrafts. In spite of official claims, the second (1969-71) and third (1973-75) reconstruction programs appear to have succeeded little in restoring the industrial base destroyed during the 1972 air war. The strengths and weaknesses of North Vietnam's industrial development to date may be summarized as follows.

First, there are a number of factors accounting for progress made during the first ten years of Communist rule: the advantage of reconstruction, the mineral base of the Northern part of the country, the rigor of Communist discipline, the priority attached to heavy

TABLE 8.4

Distribution of Labor Force
(in percentage)

Sector	1960	1965	1968	1971
Industry	7.34	8.70	10.86	12.35
Agriculture	85.32	84.20	78.60	76.81
Construction	1.84	3.03	3.93	4.07
Transport and communication	1.41	1.60	3.22	2.57
Commerce and services	3.10	2.45	3.18	4.03
Others	0.99	0.02	0.21	0.17
Total	100	100	100	100

Source: DRVN, La Republique Democratique du Vietnam (Hanoi: Editions En Langues Etrangeres, 1975), op. cit., p. 69.

industry and, last but not least, the massive flow of foreign aid from the socialist countries.

Second, the magnitude of production output indicated in this chapter may be viewed as representing the trend and the potential of the Northern economy. Reservation must be made in regard to exaggeration of these outputs, a feature of most statistical data from Communist countries. The conspicuous absence of physical-production data in absolute terms since 1964 has aggravated the statistical problem.

Third, there is also a problem of low-quality and substandard production in North Vietnam's industrial products. This problem is particularly acute in the manufacture of tools and machinery.[21] Among the factors accounting for poor quality and substandard production and maintenance, lack of incentive and poor management are perhaps the most important ones. Poor management is a broad term often used in the North Vietnamese press to refer to inefficiency, poor labor discipline, poor organization, and corruption.[22]

Fourth, the serious setback in industrial output after 1965 was due to several factors: the impact of the air attack, North Vietnam's policy of dispersion of industry, and the regime's decisive move to change its priority from industrialization to total warfare. The shifting of priority not only resulted in budgetary resources being drained for defense at the expense of economic investment, but also in the mobilization of nearly the total male labor force for war-related activities.[23]

Underlying the changing of priority was, of course, the gamble that North Vietnam would win the war and, therefore, industrialization could be postponed until the country is unified, a gamble which turned out to be profitable. The following chapter examines the prospects and problems of a unified Vietnam.

NOTES

1. When North Vietnam's policy makers discussed their industrialization plan, the most often quoted theoretical phrase to justify the state's priority given to heavy industry is from Lenin's writing: "The only possible economic base of socialism is large mechanical industry. Whoever forgets this is not a communist" (Le Duan, La Revolution Vietnamienne: Problems Fondamentaux Taches Essentielles [Hanoi: Editions En Langues Etrangeres, 1970], pp. 108-09). See also, Vo Nhan Tri, p. 375.

2. DRVN, Plan Triennal (Hanoi: Editions En Langues Etrangeres, 1959), p. 36; Vo Nhan Tri, Croissance Economique de la Republique Democratique du Vietnam (1945-65) (Hanoi: Editions En Langues Etrangeres, 1967), pp. 299, 370.

3. D. T. Truyen and P. T. Vinh, Building an Independent National Economy in Vietnam (Hanoi: Foreign Languages Publishing House, 1964), p. 80; Nhan Tri, op. cit., p. 373.

4. Le Thanh Nghi, "L'Economie du Nord Vietnam apres Trois Ans de Restauration (1973-1975), Le Courier du Viet Nam (Hanoi), no. 45, February 1976, pp. 4-5.

5. Nhan Tri, op. cit., p. 385.

6. A description of the dispersion of industry during war time is provided in Jon M. Van Dyke, North Vietnam's Strategy for Survival, (Palo Alto: Pacific Books, 1972), pp. 195-98.

7. Compared to the 1961-64 level, average investment in transport and communication during 1965-68 was four times higher, and during 1972, nearly four and one-half times higher; more than half of this investment was for repair and construction of roads and oil pipeline (DRVN, Etudes Vietnamiennes, no. 44, 1976, p. 204).

8. Thanh Nghi, op. cit., p. 4.

9. According to the Third Party Congress's guideline, electric-power development "must go one step ahead of other branches"; for different stages of electric-power development, see "Development of the Power Industry," Le Courrier du Vietnam, no. 40, September 1975, pp. 12-14, 29.

10. DRVN, Etudes Vietnamiennes, no. 44, 1976, p. 244; Le Courrier du Vietnam, no. 40, September 1975, p. 14.

11. Leon Lavallee, L'Economie du Nord Vietnam (Paris: Centre d'Etude et de Recherche Marxiste, 1971), p. 63.

12. Le Courrier du Vietnam, no. 40, September 1975, p. 14.

13. Van Dyke, op. cit., p. 199.

14. See Tran Quang Thuyet and Dao The Tuan, "Un Engrais Vert: l'azolle," in DRVN, Etudes Vietnamiennes, no. 38, 1973, pp. 127-35.

15. Lavallee, op. cit., p. 63.

16. Ibid.

17. DRVN, Etudes Vietnamiennes, no. 44, 1976, p. 245.

18. Nhan Tri, op. cit., p. 400.

19. Harvey H. Smith, et al., Area Handbook for North Vietnam (Washington, D.C.: U.S. Government Printing Office, 1967), p. 306.

20. Lavallee, op. cit., p. 63.

21. Nhan Dan (People's daily), August 11, 1968.

22. Complaints about inefficiency, mismanagement, waste, corruption, and poor organization have appeared frequently in Nhan Dan. The situation became so dangerous that it required action from the party's Central Committee (directives given by the 20th and 22d Plenums) and from the government (Resolution 19-CP, January 29, 1976 and Resolution 61-CP, April 5, 1976), calling for the improvement of "economic management" as an essential task for 1976 and 1977. For a recent and comprehensive essay on the subject, see "Ra Suc Chan Chinh To chuc San Xuat Dua Quan Ly Kinh te vao Nep va Cai Tien Mot Buoc Hoan Thanh Thang Loi Ke Hoach Nha Nuoc Nam 1976" (All-out effort to improve production organization, bringing economic management in order, and advancing one step in realizing the 1976 state plan" (editorial in Hoc Tap, no. 8, 1976, pp. 1-11).

23. On shortages of labor, see Van Dyke, op. cit., pp. 94-99, 199, 201-03.

PART V

THE END CONDITIONS: VIETNAM UNIFIED

CHAPTER

9

TOWARD ECONOMIC UNIFICATION

North Vietnam's ultimate goal of two decades of struggle (1955-75) and, to a large extent, the goal of economic development itself was to unify the country under communism.[1] It is natural that, when victory was achieved, its leaders proceeded immediately toward fulfillment of that goal.[2]

Stage one of the process is to complete unification at the state level, that is, political, administrative, and military unification. Thus a population census was taken and an administrative reorganization was carried out to elect a national assembly.*

According to new administrative division, the unified Vietnam comprises three large cities and 35 provinces, and as of February 1976 has a population of 47 million (see Table 9.1). The new General Assembly, which met in April 1976, formalized unification and decided on changes in governmental institutions and structure pending a new constitution.[3]

Efforts to unify the country were climaxed with the convening of the party's Fourth Congress in December 1976, followed by the dismantling of the National Liberation Front and absorbing of it into a loosely organized outfit, the Vietnam Fatherland Front, in February 1977.[4]

The objective of the unification at the state level was to create conditions favorable for unification in other fields,[5] especially in in the economic domain. The strength of a unified economy as seen

---

*Following a political consultative conference on unification in Saigon, November 15-21, 1975, a 492-member General Assembly was elected on April 4, 1976.

by the Communist regime and frequently spelled out by its leaders is summarized as follows:

> Our reunified Vietnam possesses a territory of more than 320,000 square kilometers, a large continental shelf, 3,200 kilometers of coasts, more than 5 million hectares of farmland and more than 15 million hectares of forests and forest land. Our country has rich natural resources, 45 million people, about 21 million workers, and a fairly large body of technical cadres and workers who have been trained in combat and production. We still have many capabilities for increasing the amount of area under cultivation in the South and the northern mountainous and hilly areas. We have built a number of important heavy industry installations in the North such as electricity, machinery, metallurgy, coal production units. The South has many capabilities for satisfactorily solving problems regarding grain, food, and raw materials for industry and export products. Both zones have large industrial complexes and harbors. Recently, we took over, almost intact, all U.S.-puppet industrial, communications and transportation installations, including a number of installations equipped with modern technical equipment.[6]

To mobilize these resources is, however, not an easy task; there are serious problems standing in the way of unifying the two economies.

PROBLEMS OF UNIFICATION

In the main, problems of economic unification stem from the vast differences between the economic systems of the two regions. Twenty years of socialist building have completely transformed the Northern economy into perhaps the world's most rigid centrally commanded economy, while a decade of massive American presence has turned the Southern economy into mainly a free-enterprise capitalistic system. Foreign resources, wartime spending, together with vast resources of the South, have created an unprecedented high standard of living for the people of the South, while a long war has exhausted the economy of the North and reduced the population's consumption to a bare subsistence level.

To the extent that real unification permits free movement between the two zones, it will for the first time expose Southern

TABLE 9.1

Vietnam: Area, Population, and Density, by Major City and Province

City/Province	Area (square kilometers)	Population[a] (thousand)	Density (persons per square kilometer)
Cities			
Hanoi capital	597	1,443.5	2,418
Ho Chi Minh	1,845	3,460.5	1,876
Haiphong	1,515	1,190.9	786
Provinces			
Lai Chau	17,408	265.6	15
Hoang Lien Son	14,125	677.2	48
Ha Tuyen	13,519	686.4	51
Bac Thai	8,615	752.9	87
Cao Lang	13,781	843.9	61
Son La	14,656	410.1	28
Vinh Phu	5,187	1,579.5	305
Ha Bac	4,708	1,466.2	311
Quang Ninh	7,076	701.8	99
Ha Son Binh	6,860	2,041.6	298
Hai Hung	2,526	1,929.9	764
Thai Binh	1,344	1,416.2	1,054
Ha Nam Ninh	3,522	2,574.6	731
Thanh Hoa	11,138	2,262.1	203
Nghe Tinh	22,380	2,704.6	121
Binh Tri Thien	19,048	1,751.8	92
Quang Nam Da Nang	11,376	1,414.4	124
Gia Lai Cong Lum	18,480	465.0	25
Nghia Binh	14,700	1,789.1	122
Dac Lac	18,300	372.7	20
Phu Khanh	9,620	1,066.2	111
Lam Dong	10,000	343.1	34
Thuan Hai	11,000	836.9	76
Song Be	9,500	561.4	59
Dong Nai	12,130	1,260.3	104
Tay Ninh	4,100	625.9	153
Long An	5,100	828.8	163
Tien Giang	2,350	1,137.2	484
Ben Tre	2,400	932.0	388
Dong Thap	3,120	991.3	318
Cuu Long	4,200	1,319.1	314
An Giang	4,140	1,361.7	329
Hau Giang	5,100	1,870.4	366
Kien Giang	6,000	834.0	139
Minh Hai	8,000	981.1	123
Total	329,466	47,149.9	143

[a]As of February 1976.

Source: Hoc Tap, no. 7, 1976, p. 88.

affluence to the North and, vice versa, will expose Northern poverty to the South. On the one hand, the strength of the Southern economy, the availability of consumer goods, the dynamic forces of a free market, prices, wages, ownership, and incentive can be expected to pose serious challenges to Northern institutions already besieged by problems of mismanagement, corruption, inefficiency, and black market. On the other hand, the weakness of the Northern economy, the rationing of goods, and the rigidity of control serve to strengthen the Southern resistance, either active or passive, to change.

The differences between the two regions have been fully recognized by the Northern regime. Its two most powerful figures have explained these differences rather comprehensively on important occasions. Truong Chinh, the National Assembly Standing Committee chairman, commented at the National Reunification Conference (November 1975):

> At present the two zones still have some differences which should be done away with step by step in the revolutionary process.
>
> Economically speaking, at present the north has two main economic sectors: The State-owned economy and the collective economy . . . the remaining individual economy is negligible, while the south has five economic sectors: State-owned economy, collective economy, joint State-private economy, private capitalist economy, and individual economy. In the north, the socialist right and collective ownership of the main means of production and the socialist relations of production enjoy absolute predominance. In the south the socialist ownership of the means of production and the socialist production relations have been established only recently. The economy of the north is planned while in the south, economic planning has just begun and on a very small scale. The wage system, the tax system, the price and monetary policies of the two zones are not yet unified.[7]

Le Duan, the party secretary general, reported to the party's Fourth Congress (December 1976):

> Thirty years of continuous and atrocious war have left throughout the country very heavy sequels in many fields. As regards economic structure there are still notable discrepancies between the two zones. In the south socialist transformation has just begun; the exploiting classes remain; the poisons of the enslaving culture and the social evils caused by U. S. neo-colonialism as well as the influence of bourgeois ideology in society remain heavy; the reactionaries are still operating against the revolution, the negative aspects of capitalism and the spontaneous character of small production are still to be overcome. This situation is confronting the revolution, especially at this initial stage, with many difficulties, and making the class struggle to solve the "Which will win" question very acute and complicated.[8]

Economic unification is, indeed, a complicated process and may only be achieved in the course of years. Even with conditions conducive to change, which existed in North Vietnam when the Communists took power in 1954 (the economy was characterized as being behind the bamboo curtain, backward agriculturally, and featuring low income, a landlord class, and nearly a century of colonialism), it took the regime more than five years (1954-60) to effectively transform the economy into socialism. Given the prevailing economic and social conditions in the South in 1975, socialist transformation there will eventually require a much longer period to be completed.

MODALITY OF UNIFICATION

Against this background, the modality for economic unification appears to include the following components:

1. Socialist transformation of the South following the Northern model with some degee of gradualism and flexibility.
2. Selective economic unification designed mainly to resume inter-regional trade and migration, under state trading and strict control, and to unify foreign trade, aid, and taxation.
3. Indicative development planning for the whole country beginning with the 1976-1980 five-year plan.

Socialist Transformation in the South

Socialist transformation is an urgent task for the South in order for it to "catch up with the North,"[9] or to arrive at an ideological balance with the North. However, a certain degree of gradualism and flexibility in the transformation is envisioned perhaps by some faction of the Politboro in view of the prevailing conditions in the South and past experience in the North.[10]

Gradualism is seen through its tolerance, at least in the initial stage, of some form of private enterprise, market prices, and profits as "economic levers"[11] in order to increase production:

> In the south, the transformation of the production relations has only just begun and remains a very heavy and complicated task . . .
>
> For some time and in certain fields of production, the economy in the south will still be composed of many sectors.[12]

Flexibility means some modification in the application of North Vietnam's socialist model to the South especially in agriculture. Modification is required both in view of past experience in the North and of the existing land-tenure system in the South.

Transformation of Agriculture

There appear to be two somewhat conflicting aspects of the process of socialist transformation of agriculture: the possibility of shortening the number of stages of collectivization, and the prospect of prolonging each stage. The seeming paradox stems from the fact that, contrary to the situation in North Vietnam in 1954, the land-tenure system in South Vietnam in 1975 was such that there was no need for the new regime to carry out land reform, and subsequently the stages of collectivization may be reduced. However, because of this reality, the new regime is deprived of a tremendously effective propaganda weapon to manipulate the masses.[13] Consequently, the progress of collectivization is expected to be slow in each stage. Furthermore, the prevailing economic conditions in the South (as already described) will definitely contribute to the Southern peasants' passive response to collectivization; thus stagnation in agriculture in the coming years appears inevitable.

The Northern model of rural transformation (as described in Chapter 4) included three main stages: land-rent reduction (1949-53), land reform (1954-56), and collectivization, which comprised the use of work-exchange teams (1956-58), low-rank cooperatives (1958-60), and high-rank cooperatives (1960-72).

Therefore, five steps were needed to change agriculture from private ownership to a system of high-rank socialist cooperatives. However, the land-tenure condition in the South in 1975 was different than the condition that had existed in the North in 1954. This is so because as early as 1956, land-reform measures were already promulgated in the South. Under Ordinance 38, the former government of the Republic of Vietnam limited land ownership to 100 hectares and imposed lower land rent. Though the 1956 program did not completely succeed, it was followed by a well-designed land-to-the-tiller program instituted in 1970.

The program, which aimed at "providing ownership to farmers actually tilling their land," was comprised of three elements: expropriating, with fair compensation, lands that are not directly tilled by landowners for free distribution to farmers; eliminating farm tenancy and abuse of accumulative land sublease by middlemen, and distributing communal riceland to farmers.[14]

The program has been considered the most successful revolutionary change in South Vietnam. By the end of 1974, 1.3 million hectares were actually distributed to more than 1 million farmers, or an average of 1.3 hectare per farmer.[15] This compares favorably with the result of land reform in the North: 810,000 hectares redistributed to 2.1 million farmers by the end of 1957, or an average of 0.4 hectares per farmer. The success of land reform in the South prior to the Communist takeover explains the total absence of the call for land-rent reform on the Communist agenda for the South. Instead, the call was for "coordinating socialist transformation with socialist building," that is, for advancing directly to the organization of work-exchange teams.

Nevertheless, experience in the North dictates that collectivization is a time-consuming process and cannot be undertaken with harsh measures:

> We must persuade, teach, and organize farmers to carry out agricultural collectivization in a gradual, positive and steady manner, thereby strengthening the alliance of workers and farmers.
>
> We must patiently mobilize farmers to take voluntary action and actively prepare the necessary conditions by stabilizing production, training management and technical cadres, consolidating party chapters, streamlining the district level, etc.[16]

Against this background, the process of socialist transformation of Southern agriculture in the coming years is likely to include five measures:

1. Formation of work-exchange teams and mutual-aid teams, beginning with work on irrigation projects.
2. Establishment of pilot cooperatives to gain experience, especially in areas where possibility for large-scale production exists (that is, in provinces where agriculture is highly mechanized).
3. Building of state farms to strengthen the state sector.
4. Gradual taking of control of the rice trade.
5. Massive doctrinal education of the masses.

Supporting all these measures is the opening up of the so-called new economic areas (NEAs) in the South.

The NEA program has a dual purpose: economically, to solve the urban-unemployment problems and increase production; politically, to provide opportunity for fast collectivization, since it is rather easy to start the work-exchange teams, to form cooperatives, and to build state farms in the newly opened areas. The program has been given highest priority among all the agricultural projects. The urban population is compelled to move either to the population-dispersion areas (those which are already cultivated or readily cultivable) or to the NEA proper (new land). These areas cluster around the provinces of Tay Ninh, Kien Tuong, Thu Dau Mot, Long Khanh, Ba Ria, Long An, My Tho, and Ben Tre. The initial stage of distribution appears to follow the former Republic of Vietnam's planned pattern for dispersion. The difference between the two schemes stems from, first, modality of dispersion (compulsary versus voluntary) and, second, presettlement preparation (almost no basic utilities made available, versus adequate infrastructure). In theory, those who go to the population-dispersion areas receive 500 square meters of land and can borrow 1,500 square meters for farming, while those who go to the NEA receive 1,000 square meters of land, and can borrow another 5,000 square meters for farming. Because of the reluctance to leave the cities and the hardship to be endured in the new land (diseases, mosquitoes, serpents), lack of infrastructure (water, health care, schools, roads, churches), and lack of food, there has been evidence of slow progress in the programs.[17]

Consolidation of State Power in the Modern Sector

In industry, the prospect for socialist transformation appears to be better than in agriculture. For one thing, the public-enterprise sector in the former Republic of Vietnam (some 50 companies) already covers a wide range of activities: electricity, communications, bus, rail, and airline service, sugar, cement, paper, textile, agricultural machinery, and construction.[18] In addition, large corporations left behind either by foreign business or by refugees who left the country in 1975 can easily be nationalized.[19]

In regard to the small-business sector, the socialist-transformation process would be more gradual. For a period of three to five years, it would not eliminate small business, but would "use, restrict and transform the private capitalist through the establishment of joint state-private enterprises,"[20] and at the same time control and guide the private enterprises that still are allowed to exist in order for their operation to conform to "a planned socialist economy" and one "serving the interest of socialism."[21] A certain amount of economic levers or incentives (prices, wages, profits, capitalist management) will be tolerated with a view to increasing production. The incentive-for-production position is explained this way by a North Vietnamese economist:

> In South Vietnam, socialist reform and the reorganization of social production will be carried out under the conditions of an economy which still consists of many different segments; economic management and planning will be based on heightening the dominant and guiding role played by the state operated economy in accordance with the state plan while making use of the commodity-money relationships, the relationships between the supply of materials and the purchasing of products, and the levers of prices, credit, wages, bonuses, and profits to stimulate the development of production, insure that the economy develops within the socialist orb, and constantly consolidates socialist production relations.[22]

Le Duan describes the position as follows:

> We must attach importance to value and the law of value which still exist objectively in the socialist society; we must make flexible use of the law of value and many other economic levers to strengthen economic and financial management, encourage labour and promote the increase of labour productivity, reduce the consumption of labour, materials and equipment per unit of product, and ensure the production of the greatest amount of use-values with the lowest expenditures. In this spirit, we

> must make proper use of the market, prices, wages, and credits to improve the quality of the plan, better serve the working out and implementation of the plan and, in a proper measure, to complement the plan.[23]

Selective Unification

Side by side with the transformation of the South to reach the socialist balance with the North, selective-unification measures are being taken to lift the North to economic balance with the South.

Resumption of Interregional Trade

As emphasized throughout this book, the most dramatic change in the Northern economic life after the victory over the South has to be its access to the Southern food supply. While the political leaders prefer to elaborate generally on the strength of a unified economy as already noted, North Vietnam's economists were more explicit on the food point:

> In the past, North Vietnam encountered difficulties trying to resolve the grain problem by itself. Today, with the capability of the entire country we will concentrate our efforts on rapidly increasing the amount of area under the cultivation of rice and rapidly increasing rice yields while intensively cultivating and increasing the amount of area under the cultivation of subsidiary food crops, especially corn, potatoes, and cassava; in a short amount of time, we will be producing enough grain to meet the needs of the people, develop livestock production, and store grain.[24]

It was officially reported in February 1976 that "economic trade between the South and the North has begun, thereby helping to develop the great potentials of the economy of the entire country."[25] The new interregional trade is expected to be beneficial to both zones:

> The North-South commercial relation which has already begun, will play a more and more important role in the economic development of the country. Coal, phosphate fertilizer, certain mechanical products of the North are precious to the South which, in turn,

> supplies foods [and other] consumer goods to the North.[26]

There have been reports that shipments of rice and other foodstuffs from Saigon to Haiphong had already begun in 1975.[27] The 1976 plan called for smooth circulation of the North-South flow of goods, and the five-year plan (1976-80) envisaged the complete modernization of the trans-Vietnam rail line.[28]

Given the fact that trade flow will be selective and strictly in the hands of the state monopoly, railroad transportation offers the best way for efficient control of the traffic. Consequently, throughout 1975 and 1976, an all-out effort was made to reconstruct the Hanoi-Saigon railway, involving, in addition to hundreds of thousands of workers, 30 enterprises, 70,000 technicians, and army units.[29] By the end of 1976, the task was considered completed, and the railway was ready to welcome the party's Fourth Congress:

> One of the finest flowers we wish to present to our congress is the reunification railway, linking Hanoi, our capital, to Ho Chi Minh City, which was interrupted for over 30 years and which has been rehabilitated ahead of schedule. These new factors will certainly lead to more profound changes and greater successes in the coming period.[30]

Unifying Financial Management

Unifying financial management, which means total control of South Vietnam's Central Bank, the commercial banking sector, other credit institutions, and the exchange of currency, comprises the first set of financial policies for the South.

The taking over of the Central Bank and commercial banks was the most important step to gain total control of foreign exchange (see Appendix 9B), foreign trade, domestic trade, and purchasing power of the population.*

In the Communist history of revolution, the failure of the revolutionary army to take over the former bank of Indochina, was viewed in both Hanoi and Saigon as one of the major weaknesses of

---

*By substantially reducing the per-capita money supply through currency-exchange measures, the purchasing power of the population was sharply cut back. Under new currency-exchange measures, strict limit was put on the amount of old currency each person was allowed to exchange for new currency.

the August 1945 revolution.[31] This fact explains the speed with which the new regime seized control and management of South Vietnam's Central Bank upon the occupation of Saigon. There is another parallel in the two cases: the issue of new currency.

Five months after the August 1945 revolution, the Communist regime issued its currency, the dong, in North Vietnam.[32] And five months after the April 1975 victory, the new regime also issued the dong in South Vietnam. The 1975 exchange rate was one dong to 500 piasters. In theory, all the piasters were to be exchangeable for new money, but in reality each family having less than 1 million piasters was allowed to exchange 100,000 piasters (roughly U.S. $1,000 in the spring of 1975) for 400 dong and to deposit the rest with the Central Bank. Small businesses with certified licenses can exchange an additional 300,000 piasters for another 600 dong (see Appendix 9A).

The new currency and currency exchange was designed only for the South, where Northern money is not permitted to circulate as clearly stipulated by the government:

> Northern money may not be circulated on the market. Cadres and employees having Northern money should deposit it with the national bank in the area of their assignment . . . .
>
> The Vietnam National Bank of the South will accept only those amounts of Northern money of people having certificates of authorization to bring Northern money into the South and sums certified by agencies as money brought into the South previously that the possessor had not wanted to exchange.[33]

The simultaneous existence of two currencies in one country dramatizes the selectivity of economic unification, and indicates the slow-moving process of economic integration of the two economies. In the first stage of this process, a unified economic-development plan for the entire country appears to be the symbol of economic unification, to be completed in a still uncertain period of time.

CONCLUSION

The fall of South Vietnam in the spring of 1975 opened the way for the unification of the country under the Communist regime—the goal of 20 years of struggle.

Unification began with political and administrative measures to bring the South firmly under the control of the Communist party.

This is to be followed by complete unification in other fields, especially the economy. Economic unification is, however, a very difficult and time-consuming process which may never be completed, given the structural differences between the two economic systems resulting from 20 years of separation and of socialist building in the North as well as capitalist orientation in the South.

The first stage of the unification process as it is now unfolding, may be characterized by selective unification measures. Among them, the most important are the resumption of interregional trade, the taking over of the Southern financial system, and the socialist transformation of the South.

Given the prevailing conditions in the South, however, the application of the Northern socialist model is expected to progress with a certain degree of flexibility and gradualism, at least for a few years. In the meantime, it seems rational to the Communist regime that the symbol of economic unification can be best illustrated by the launching of a unified economic-development plan for the entire country. The plan is analyzed in the next chapter.

APPENDIX A: GUIDELINES FOR NEW CURRENCY EXCHANGES*

1. All sums of money the people currently have can be exchanged for the new currency. This represents the people's right of ownership.

2. As an immediate step to assure the activities, production and enterprise of families, in accordance with the regulations that have already been announced, the Vietnam National Bank allows the immediate exchange of the following necessary amounts:

> a. Families having 100,000 piasters [equivalent to about U.S. $100 in early spring, 1975] and less can exchange the actual amounts they have.
>
> b. Families having more than 100,000 piasters can exchange up to this amount right now.
>
> c. Enterprise families can exchange money according to the regulations that have already been announced. The transactions office of the bank is

---

*Excerpted from article in Saigon Giai Phong, September 24, 1975, translated by U.S. Department of Commerce, Joint Publications Service, <u>Translations on North Vietnam</u>, no. 1737, December 1, 1975, pp. 32-33.

assigned the responsibility for exchanging money up to the maximum of 200,000 piasters; the bank branches in precincts and districts up to 500,000 piasters; and higher amounts by decision of provincial and city branches.

3. In the case of people having from over 100,000 to 1 million piasters, after the exchange of the amounts necessary for their needs in everyday life, production and enterprise, if any, the remaining amounts will be put in savings accounts.

These savings will get dividends for either timeless or time-limit savings, which they will choose themselves. Timeless savings yield lower dividends than time-limit savings, but they can be withdrawn anytime. Time-limit savings yield higher dividends but cannot be withdrawn before maturity, except in case of real necessity they can be withdrawn before the maturity date.

4. In the case of people having sums of money higher than 1 million piasters which they have also declared, after the necessary exchanges for production and enterprise needs, their money will be deposited and yield dividends in accordance with the deposits policy and can be withdrawn to satisfy production and enterprise needs in accordance with the normal procedures of the bank.

5. The savings and deposits made with the Vietnam National Bank today are different from those made with the private banks and banks of the old Saigon puppet administration. The Vietnam National Bank assumes full responsibility toward its customers for these new savings and deposits.

APPENDIX B: VIETNAM NATIONAL BANK'S FOREIGN EXCHANGE CONTROL*

(Measure Dated September 10, 1975)

I. Holding, circulating, trading, pledging, and conveying foreign exchange in any form within the territory of the Republic of South Vietnam are all prohibited, excluding cases authorized by the Vietnam National Bank.

---

*Saigon Giai Phong, September 14, 1975, pp. 1, 4, translated by U.S. Department of Commerce, Joint Publications Service, Translations on North Vietnam, no. 1734, November 24, 1975, pp. 36-37.

a. Whoever has foreign exchange, regardless of source, such as foreign silver certificates, foreign coins, all types of share certificates and shares, all instruments of payment inscribed in foreign currency (such as all types of checks, postal orders, warrants, commercial paper, negotiable letters of credit), and so forth must sell it to or deposit it with the bank assigned jurisdiction within 10 days following acquisition of the foreign exchange.

b. Precious metals, precious stones, and pearls when exported from or imported into the Republic of South Vietnam take on the nature of foreign exchange within the scope of state control and permission must be obtained from the Vietnam National Bank for export or import.

c. Those found by the Vietnam National Bank to have the status of nonresidents may be authorized by the National Bank to import, sell, deposit, or hold foreign exchange in the Republic of South Vietnam without limit on quantity. These persons must file a declaration with and request permission from the Vietnam National Bank for the sum of foreign exchange on hand in South Vietnam or that will be imported.

The foreign exchange of nonresidents may be sold to the bank assigned jurisdiction in order to obtain Vietnamese currency to expend in South Vietnam, and if not all is expended that remaining may be reexchanged for foreign currency when leaving Vietnam.

II. Procedures for declaring, sale, or deposit of foreign exchange mentioned above as stipulated by the Vietnam National Bank are:

a. Foreign silver certificates and coins that are exchangeable (British coins, U.S. coins, French francs, West German marks, Swiss francs, yen, Hong Kong dollars, Singapore dollars, Malaysian dollars, and so forth) will be purchased at the exchange rate announced daily by the Vietnam National Bank.

b. As for all types of share certificates, shares, and all types of instruments of payment inscribed in foreign currency, depending on the case, the Vietnam National Bank will authorize the bank of jurisdiction to make immediate purchase, to receive for resale as the agent, or hold as the agent for the person having the foreign exchange.

c. Vietnam Thuong Tin has been entrusted by the Vietnam National Bank with receiving declarations of, trading in, and receiving and holding foreign exchange covered in Article I and a and b of this Article II.

d. As for precious metals (gold, silver, and so forth), precious stones, and pearls, the Vietnam National Bank will itself take charge of controlling trading, holding, and conveying them within all territory of the Republic of South Vietnam in accordance with a separate notice.

III. Those who presently have foreign exchange must, between now and 15 October 1975, declare it and sell it to or deposit it with the Republic of South Vietnam in accordance with the above stipulations; the source of this foreign exchange need not be noted.

However, for foreign exchange received or brought into the Republic of South Vietnam legally within 10 days of 15 October 1975, the deadline for declaring and depositing or selling the foreign exchange will be 10 days counting from the day of receipt or import of the foreign exchange into the Republic of South Vietnam.

IV. Effective the day of promulgation of this notice, anyone circulating, trading, pledging, conveying, failing to declare, or inaccurately declaring in contravention of this notice must bear full responsibility at revolutionary law for violation of state foreign exchange control regulations.

V. Previous stipulations on foreign exchange control that conflict with this notice all are abolished.

10 September 1975
Governor, Vietnam National Bank

NOTES

1. The goal of national unification was laid down by Ho Chi Minh after the division of the country in 1954. When Ho died, unification was made his ultimate wish in his testament. See Ho Chi Minh, Selected Writings (1920-1969) (Hanoi: Foreign Languages Publishing House, 1973), pp. 361-62; for statements in early years (1954-56), see pp. 187, 190-91, 193.

2. After the conquest of South Vietnam, Le Duan declared in June 1975: "We are very proud to have thoroughly carried out these recommendations (American defeat and national unification) of

President Ho Chi Minh in his sacred testament." See Le Duan, "Forward to the Future," Le Courrier du Vietnam, no. 37, June 1975, p. 4. See also Truong Chinh, "Toward Completion of National Reunification: The Substance, Objectives and Urgent Problems To Be Solved," Le Courrier du Vietnam, no. 43, December 1975, pp. 3-7.

3. The Assembly met in Hanoi from June 24 to July 3, 1976 to formalize unification. For details, see Hoc Tap, no. 7, 1976, pp. 63-68.

4. On the liquidation of the front, Le Duan explained:

> In the recent past, the Vietnam Fatherland Front and the South Vietnam National Front for the Liberation have fulfilled their glorious task of mobilizing and rallying all strata of people . . . to complete the liberation of the south, thus creating a tremendous force to defeat the enemy . . . . Now that the task of the national people's democratic revolution has been completed throughout the country, now that we are switching to the socialist revolution, the character, tasks, composition, as well as the activities of the front also see major changes. Of late, the consultative conference between the Central Committee of the National Front for the Liberation of South Vietnam, and the Central Committee of the Alliance of National, Democratic, and Peace Forces of Vietnam has decided to convene a conference for the founding of a single national front for the whole country.

This appeared in Le Duan, "Report to the Fourth Congress, Vietnam Workers Party," December 15, 1976, broadcast by Radio Hanoi, December 16, 1976, and published in U.S. Foreign Broadcast Information Service, Asia and the Pacific (hereafter cited as APA) 4, no. 247, supp. 48 (December 22, 1976): 82.

5. Truong Chinh, "Cuoc Tong Tuyen Cu 25-4-1976 Danh Dau Buoc Thang Loi Quyet Dinh Cua Nhan Dan Ta Tren con Duong Nhat Nuoc Nha Ve Mat Nha Nuoc" (The April 25, 1976 general election has marked a decisive victory of our people on the way to unify the country at the state level) Hoc Tap, no. 7, 1976, p. 17.

6. "Endeavoring to Successfully Implement the 1976 State Plan and Advance Our National Economy Toward Large Scale Socialist Production," Hoc Tap, no. 2, 1976, pp. 1-13, translated by the U.S. Department of Commerce, Joint Publications Service, Translations on North Vietnam (hereafter cited as TNVN), no. 1784, March 22, 1976, p. 4.

7. Truong Chinh, "Toward Completion of National Reunification: The Substance, Objectives and Urgent Problems to be Solved," op. cit., p. 4.

8. Duan, "Forward to the Future," op. cit., p. 21.

9. Truong Chinh, "Toward Completion of National Reunification: The Substance, Objectives and Urgent Problems To Be Solved," op. cit.

10. See Hoang Van Chi, From Colonialism to Communism, (New York: Praeger, 1968), especially pp. 209-39. On the experience of the North, one of North Vietnam's leading economists has this to say:

> In its 20 years of economic reform and construction, North Vietnam has gained much experience in planning, organizing and managing the economy, training and utilizing technical cadres and management cadres, etc. These experiences gained in the past, both successful and unsuccessful ones, are extremely valuable for socialist construction in South Vietnam.

See Che Viet Tan, "Developing the Combined Strength of our Unified Economy," Hoc Tap, no. 12, 1975, TNVN, no. 1770, p. 13.

11. "Endeavoring to Successfully Implement the 1976 State Plan and Advance Our National Economy Toward Large Scale Socialist Production," op. cit.

12. Duan, "Report to the Fourth Congress, Vietnam Workers Party," op. cit., p. 53.

13. On the effectiveness of mass manipulation through land reform in North Vietnam, see Hoang Van Chi, From Colonialism to Communism (New York: Praeger, 1964), pp. 163-68.

14. Vietnam Council on Foreign Relations, The Vietnamese Land Reform Law (Saigon: Saigon Publishers, 1971), pp. 7-14.

15. Vietnam Bulletin (Embassy of the Republic of Vietnam, Washington, D. C.) 9, no. 6, (April 1974): 3.

16. Ho Liem, "South Vietnam in the New Stage of the Revolution," Hoc Tap, no. 12, 1975 (TNVN, no. 1770, p. 49).

17. See, for example, Andre Gelinas, "Life in the New Vietnam," Washington Post, March 13, 1977, pp. C1 and C4.

18. For information on the size of state enterprises, see Joint Development Group, The Postwar Development of the Republic of Vietnam (New York: Praeger, 1970), Table 9. 3, p. 260.

19. See Liem, op. cit., p. 48.

20. Gradualism permits the development of the positive aspects and the limitation of the negative aspects of the capitalist economy

while at the same time "insuring to the workers of appropriate positions in private enterprise" (ibid).

21. Duan, "Report to the Fourth Congress, Vietnam Workers Party," op. cit., p. 53.

22. Viet Tan, op. cit., p. 25.

23. Duan, "Report to the Fourth Congress, Vietnam Workers Party," op. cit., p. 53.

24. Viet Tan, op. cit., pp. 20-21.

25. "Endeavoring to Successfully Implement the 1976 State Plan and Advance Our National Economy Toward Large Scale Socialist Production," op. cit.

26. "Sud Vietnam 1976" (an interview with Nguyen Khac Vien after his tour of the South), Le Courrier du Vietnam, no. 47, April 1976, p. 8.

27. See Douglas Pike, "The Democratic Republic of Vietnam," in Yearbook on International Communist Affairs, 1976 (Stanford, Calif.: Hoover Institutions, Stanford University, 1976), p. 400.

28. Duan, "Report to the Fourth Congress, Vietnam Workers Party," op. cit., p. 40.

29. "The Hanoi-Saigon Reunification Railway Line," Le Courrier du Vietnam (English ed.), no. 43, December 1975, p. 7.

30. Duan, "Report to the Fourth Congress, Vietnam Workers Party," op. cit., p. 19.

31. Doan Trong Truyen and Pham Thanh Vinh, Building an Independent National Economy in Vietnam (Hanoi: Foreign Languages Publishing House, 1964), p. 20.

32. Ibid., p. 21.

33. "Vietnam National Bank's Policy, September 12, 1975," Saigon Giai Phong (Saigon liberated), September 14, 1975, p. 4.

CHAPTER

10

PROSPECTS FOR THE FUTURE: THE FIRST NATIONAL PLAN (1976–80)

To celebrate the conclusion of 16 years of the party's Third Congress and to celebrate the opening of the Fourth Congress, assembled in the capital of Hanoi in December 1976, a five-year development plan (1976-80) was presented. It is to be noted that the term "plan" used by the North Vietnamese in this context is rather loose, since in essence, the so-called 1976-80 plan is merely a compilation of guidelines, orientation, economic tasks, and a number of indicative targets for different sectors.

The 1976-1980 plan, originally intended to be North Vietnam's third five-year plan, was revised to become the first five-year plan covering the whole country. The decision to integrate the South Vietnamese economy into a broad national plan appears to be the result of a crucial resolution of the party's 24th Plenum held in the late summer of 1975.[1] The resolution called for the speeding up of national unification and of socialist transformation of the country:

> The strategic task of our country's revolution in the next stage is: completing the reunification of the country and rapidly, strongly, and steadily advancing the entire country to socialism. North Vietnam must continue to accelerate socialist construction and perfect the socialist production relations; South Vietnam must simultaneously carry out socialist reform and socialist construction.[2]

Although the original 1976-80 five-year plan has never been revealed, it seems highly possible that it was based mainly on North Vietnam's second five-year plan, which had been prepared exactly a decade earlier (1966) but had to be abandoned in the face of the air war. In fact, North Vietnam's official economic reports in 1975 insisted that the 1965 (prewar) level of population had been restored. It follows that North Vietnam could start out (in 1976) with the state of the economy as it existed in 1965.

The current five-year plan is to be only the first step in North Vietnam's long-term development prospective: the construction of a modern socialist state within 15 to 20 years. According to the party's guidelines, the economic-construction task in the future is as follows:

> To give priority to the rational development of heavy industry on the basis of agricultural and light industrial development; correctly coordinating industry and agriculture; simultaneously carrying out central economic construction and local economic development; coordinating the economy with the defense of the nation and widening economic relations with socialist and other countries . . . and endeavoring to virtually complete the construction of the material and technical base of socialism in the country in the space of 15 to 20 years.[3]

The plan was late in starting and so far there has been no comprehensive and systematic official plan released. On December 16, 1976, the plan was partially described in a political report by the party Central Committee, read by Le Duan, to the party's Fourth Congress. The analysis of the plan in this chapter is based mainly on that report.[4]

OBJECTIVES

The broad objectives of the plan are to:

> 1. Highly concentrate the forces of the whole country, of all branches and echelons, to achieve a leaping development of agriculture; strive to step up forestry and fishery; vigorously develop light industry and food industry (including also handicrafts and small industries) with the aim of steadily solving the needs of the whole country in food, foodstuffs and an important part of the common consumer goods; to improve one step further the material and cultural life of the people, and create accumulation for socialist industrialization.
>
> 2. Turn into full account the existing capacities and build many new bases of heavy industry, especially the engineering industry, in order to serve first of all

agriculture, forestry, fishery and light industry, and make provision for technical equipment for the following period. Actively expand communications and transport; rapidly increase the capacities of capital construction; push ahead scientific and technological work, make preparations in all fields for large-scale construction in the future long-term plans.

3. Use all the social work force; organize and manage well the labour force in order to markedly increase social labour productivity; take the initial steps in establishing an industrial-agricultural economic structure, combining central with local economies; step by step build the district into an agro-industrial economic unit; combine economy with national defense, build an all-people national defense.

4. Basically achieve socialist transformation in the south; consolidate and perfect the socialist production relations in the north; vigorously improve trade, prices, finance and banking operations.

5. Rapidly increase the sources of export, first of all agricultural and light industry products; broaden economic relations with foreign countries.

6. Strive to develop education, culture and health work; carry out an educational reform; push ahead the training of cadres and workers; do away with the sequels of war and neocolonialism on the social plane, and

7. Effect a deep change in economic organization and management, set up a new sysetm of economic management on a national scale. [5]

Among these general objectives, numbers 1, 4, 5, and 6 appear to be more relevant for the South, while numbers 2 and 7 are for the North, and number 3 is for the whole country.

The plan's implication is that the South is to return to its traditional role, that is, to supply the food requirement for the nation and for export. In addition, because the South is endowed with a modern light-industrial base, it is expected to concentrate on production of consumer goods (especially food products), and on light industry and machinery repairs. North Vietnam's role, in addition to leading the country toward communism, will be to accelerate the development of heavy industry, especially electricity, coal, steel, and machinery.

TABLE 10.1

Vietnam Unified: First Five-Year Plan (1976-80)—
Specific Targets and Estimated Rates of Growth, Compared with 1964-65 and 1974-75

Sector	1964-65 Level			1974-75 Level			Targets for 1980	Estimated Growth (in percentage)	
	South[a]	North	Total Vietnam	South[a]	North	Total Vietnam	Total Vietnam	Total 1976-80	Average Annual
Agriculture									
Food (million tons)	6.1	6.7	12.8	8.7	6.4	15.1	21.0	39	7.8
Rice (unprocessed, million tons)	4.8	4.3	9.1	7.2	4.9	12.1	—	—	—
Area under cultivation (million hectares)	2.9	3.3	6.2	3.1	2.9[b]	6.0	7.0	16	3.2
Cultivated (million hectares)	2.9	3.3	6.2	3.1	2.9	6.0	7.0	16	3.2
Rice land (million hectares)	2.6	2.4	5.0	2.9	2.1	5.0	6.0	20	4.0
To be reclaimed during 1976-80 plan	—	—	—	—	—	—	1.0	—	—
Newly planted forest area (million hectares)	1.1[c]	2.0[d]	3.1	1.1	2.0[d]	3.1	4.3	38	7.6
Seafood (thousand tons)	375	85[c]	460	728	100[d]	828	1,000	20	4.0
Animal husbandry (million head of pigs)	3.5	4.5[c]	8.0	5.1	6.4	11.5	6.5	43	8.6
Industry									
Coal (million tons)	—	4.2	4.2	—	4.6	4.6	10	117	23.4
Electricity (million kilowatt hours)	0.6	0.6	1.2	1.5	1.3	2.8	5	78	15.6
Cement (thousand tons)	189	573	762	295	390	685	2,000	191	38.2
Steel (thousand tons)	—	100[b]	100	—	—	—	300	—	—
Engineering output (percent increase over 1975 level)	—	—	—	—	—	—	150	150	30.0
Chemical fertilizer (thousand tons)	—	144	144	—	382	382	1,300	240	48.0
Textiles (million meters of fabric)	118	100	218	83	88	171	450	163	32.6
Paper (thousand tons)	19	24	43	49	12	61	130	213	42.6
Building Construction									
Housing (million square meters of floor space)	—	—	—	—	1.4	—	14	—	—

[a]1974-75 (pre-Communist rule) figures.

[b]1973 figures; because of adverse weather conditions, actual area under cultivation in 1975 was smaller.

[c]1963 figures.

[d]Estimated.

Sources: For 1965 and 1975: Tables 1.1, 7.5, 8.3; Republic of Vietnam, National Institute of Statistics, Vietnam Statistical Yearbook, vol. 18, 1972, pp. 29-328, 187-218, vol. 20, 1975 (unpublished). For 1980 targets: Le Duan, "Report to the Fourth Congress, Vietnam Workers Party," December 15, 1976, broadcast by Radio Hanoi, December 16, 1976, and published in U.S. Foreign Broadcast Service, Asia and the Pacific 4, no. 247, suppl. 48, p. 33, December 22, 1976; Doan Ket (Unity), no. 201, November 11, 1976.

SPECIFIC TARGETS

Specific targets for the plan as given in Le Duan's report are arbitrarily summarized in Table 10. 1. In order to provide some base for comparison and to estimate growth rates for each sector, data for 1965 (the last year of North Vietnam's first five-year plan) and for 1975 (the base year for the current plan) are also presented.

Agriculture

Food Grains

The planners envisaged that by 1980 the country will have produced a total of 21 million tons of food grains, implying an estimated total growth of 39 percent for the whole plan, or a 7. 8 percent average per annum. It is not known, however, how much of the total food-grain target is set for rice production. It is possible that the planners may have envisaged a somewhat declining relative share of rice in total food-grain output. With a hypothetical annual growth rate of about 6 percent, the 1980 rice output will be around 16 million tons, accounting for 76 percent of total food grains as compared with 80 percent in 1975. In light of the first five-year plan, the 7. 8 percent annual growth projected appears to be more realistic; the first five-year plan had assigned a 10 percent annual growth for food grains, and the original target was later revised to 6. 5 percent.[6]

Arable Land

In order to increase the production of food grains, one million hectares of arable land, apparently rice land, will be put under cultivation. The increase should come mainly from South Vietnam where between 2 and 2. 5 million hectares can still be opened up for farming; cultivable land in the North has long been exhausted.

Forest

To enhance the prospect for large exports of timber, the plan calls for increasing newly planted area by 1. 2 million hectares, pushing the amount of forest land exploitable to about 4. 3 million hectares or roughly one-third of the country's total land under forest cultivation.[7] Most of the expansion will take place in the highlands of North and Central Vietnam and in the U-Minh area (South Vietnam's southernmost province).

Fishing and Animal Husbandry

In fishing and animal husbandry, targets are set to increase the sea-fish catch to 1 million tons annually and the hog population to 16.5 million heads by 1980, implying annual average growth rates of 4 percent and 8.6 percent, respectively. As is evident from Table 10.1, the South will be the region mainly responsible for the growth of fishing and animal husbandry.

Industry

Coal

A target of 10 million tons of coal output in the plan's terminal year is called for, a level which is twice as large as that projected for the first five-year plan (5 million tons), but implying almost the same rate of growth: double the base year's output by the final year of the plan. Coal-output expansion will come mainly from North Vietnam, with South Vietnam playing a minor role, in view of the prospect for revitalizing the Nông Sòn mine in the Quảng Đà area.

Electricity

Five billion kilowatt hours of electric power is expected to be generated in 1980, implying a 79 percent increase over the 1975 level, or 15.8 percent annually. Given the almost equal share of total electric output for the two zones in previous years (1965 and 1975 as shown in Table 10.1), it is evident that the source of growth in the power sector will be coming from both regions, that is, increasing by about 1.2 billion kilowatt hours each, which is about twice the level projected for the first five-year plan (660 million kilowatt hours).

Cement

An ambitious target of 2 million tons of cement, nearly triple the 1975 level, is set for 1980, requiring an average annual growth rate of 38.2 percent. Traditionally, cement production was the monopoly of North Vietnam. However, since partition South Vietnam has produced its own cement and in 1975 its cement production was nearly 300,000 tons, with capacity expected to be doubled within a few years. It seems uncertain, nevertheless, whether the plan envisaged any cement production in the South.

Steel

Similar to the first five-year plan, an ambitious target is projected for the steel insustry. Assuming that the Thai Nguyen steel plant was restored in 1975 as reported, and output was around 50,000 tons of iron ingots, the 300,000-ton goal for 1980 implies high growth: a sixfold increase during the plan.

Engineering Output

The production of tools and machines in 1980 is expected to be two and a half times that of 1975, implying a total increase of 150 percent or 30 percent annually. As the engineering sector is most crucial branch of the socialist heavy industry, its growth is expected to come mainly from North Vietnam.

Fertilizer

Fertilizer production appears highest on the list of priority in terms of growth rates. The 1970 production target of 1.3 million tons of chemical fertilizer implies a total growth of 240 percent or a 48 percent average annually. The large expansion of fertilizer production, mainly in the North, is called for to boost rice production in the South in order to support the fast-growing population in both zones.

Consumer Goods

Among the consumer goods, specific targets are set for textiles and paper production, apparently to meet the two goals the leaders consider most important in addition to food supply: the need for clothing and for socialist education. Thus, 450 million meters of fabric and 130,000 tons of paper are projected for 1980, implying annual growth rates of 32.6 and 42.6 percent, respectively.

DEVELOPMENT POLICIES AND PROGRAMS

Agriculture

The overriding policies in agricultural development appear to be designed to divide the country into agricultural zones, specializing in the production of different food crops and industrial crops;[8] and to push the small farms into "large-scale socialist production."[9] In North Vietnam, large scale socialist production implies the expansion of large-scale state farms and the merging of individual cooperatives into bigger ones; in South Vietnam, it implies the drive toward collectivization.

Measures to increase agricultural output include expansion of farm land through the creation of the so-called new economic areas, mainly in the south,[10] increase in irrigation and water conservation, increase in fertilizer application, intensive farming, agricultural mechanization, and redistribution of labor force between urban and rural areas and between North and South. The latter, though not spelled out in the plan, appears to be of paramount importance on grounds of availability of arable land.[11]

Light Industry

Development policy for light industry (mainly in South Vietnam) includes the emphasis on food processing for domestic consumption and export, and on basic consumer goods, mainly for domestic consumption.

Food processing appears to resemble the pattern of production in South Vietnam in prewar Communist years and includes the milling of rice and wheat flour, processing of marine products, production of sugar and salt, and the processing of vegetables and fruits, drinking water, fruit juices, beer, and mineral water.[12]

The task of North and South in light industry is summarized as follows:

> With the unification of planning throughout the country, light industry and the food products industry of the two zones of our country can coordinate and supplement one another very smoothly. The North has important installations producing cloth and sugar, processing grain and food products, making pottery and glass, and so forth which met approximately 80 percent of the needs for consumer goods during the years of the war. In light industry, the South has such high capacity sectors as cloth, silk, resin, plastics, soap, and so forth; in the food products industry, there are sugar, milk processing, beer, soft drinks, flour, MSG, tobacco, and so forth; some sectors have reached a rather high level of technical development. During the next several years, the light industry and food industry of our country can develop in a diversified manner with many different products, thereby helping resolve the problems of food, clothing and consumer goods for the people and some for exportation.[13]

Heavy Industry

Consistent with the Marxist-Leninist ideology, construction of heavy industry lies at the heart of economic-development policy, as was the case in the first five-year plan. Within heavy industry, priority appears to be attached to mechanical-engineering production, electricity, coal, and fertilizer. The emphasis in engineering output is explicitly spelled out as follows:

> Holding the key position in the industrial system, the engineering industry must become a prime concern in the whole plan for development of the national economy. It must be rapidly built into a major and powerful branch in order to supply step by step machinery, machine tools, accessories, spare parts and whole sets of equipment to the various branches of the national economy . . .[14]

Regarding electricity development, hydroelectric power receives special attention in the plan:

> With regard to electricity, the basic orientation is to combine hydroelectric power with thermal power; attach utmost importance to the former; combine large, medium and small hydroelectric projects. Build an electric grid proportionate to the sources of electric energy. In this five-year plan, we shall expand some old power stations and build a number of new ones. Attention shall be given to building many small hydroelectric stations in the midlands and the mountain regions, especially in the Central Highlands.[15]

In North Vietnam, electric-development programs will include construction of the Uong Bi plant (phase 4 of the programs), the Dap Cau plant (phase 1), and the Pha Lai plant (phase 1); in South Vietnam, the strengthening of the Danhim hydroelectric-power plant and the Danang thermal-power plant.[16]

Totally new compared with the first five year plan, the new plan calls for manufacturing and assembling of electronic equipment and exploration of minerals and oil and natural gas. It suggests the creation of "conditions for rapidly building an oil industry and envisions the construction of the first oil refinery".[17]

Infrastructure

Among important projects mentioned in the plan are the rebuilding of the capital of Hanoi to make it "grander, bigger, and finer," modernization of the trans-Vietnam (Hanoi-Ho Chi Minh city) railroad, mechanization of ports and cargo handling and of river transport, and building (for the first time in North Vietnam's development program) of civil aviation into the transport branch.

Other Development Measures

Domestic Trade

State retail trade is to be improved in the North and state wholesale trade is to be created and expanded in the South:

> The retail network must be organized rationally, widely and strive soon to do away with too long queues for the buyers in front of the department stores, unaccountable weighing, measuring and counting, and step by step build a civilized trade service, build an attitude of good service to customers and high proficiency for all cadres and personnel of the trade service, worthy of being enterprising suppliers and devoted house-keepers of the people; resolutely combat graft and overbearing . . .
>
> In the south we must rapidly broaden the state-run trade service and the marketing cooperatives, arrange in a rational way the wholesale and retail networks, and see to it that the state-run service not only has complete control over the wholesale trade but also can manipulate the retail trade and manage the market and prices, thus creating the fundamental conditions to unify the markets in the whole country . . .[18]

International Trade

The plan calls for increasing international trade and cooperation with socialist and nonsocialist countries as a means to secure modern technology in order to help the economy advance from small- to large-scale socialist production:

> The aim of exporting and importing is to secure the world's modern technology, to make production and

> labor achieve higher economic results in our country, thus contributing to the growth of accumulation and the improvement of our people's life. Exports should be further impelled to stimulate production, raise the technical level, promote the division of labor and create conditions for broadening imports. Imports must satisfy the requirements of the scientific-technological revolution and the building of the material-technical basis of socialism; they must help increase labor productivity in our country . . .[19]

Fiscal and Monetary Policy

According to the plan, the objective of fiscal policy cannot be confined to fiscal neutrality aimed at merely "increasing revenues, reducing expenditures and balancing the budget." Policy should also firmly control the "fixed assets and circulating assets" (wealth and income) in order to finance economic development, to allocate among consumption, accumulation (investment), and defense. The plan specifically calls for a "correct tax policy" in the South to help regulate the source of income. Stripped of rhetoric, this implies mainly tax in kind in agriculture (that is, rice).

The plan stipulates the eventual unification of the budgetary system and of markets and prices. It specifies that the state banking system should

> take part in working out a plan for economic development, . . supply credit capital for production and business activities, . . . use money to supervise the productivity, efficacity, and quality of these operations, . . . become clearing centers of all production and business operations, effectively manage cash money and regulate in time monetary circulation in society . . .[20]

PROSPECTS FOR FINANCING THE PLAN

There are no indications in the plan as to how much is total investment required and how it is going to be financed. The planners appear to count heavily on the agricultural resources and petroleum prospects of the South, the large labor force, rigorous socialist control to transfer resources from agriculture to industry, and especially on foreign aid.

If the pattern of financing reconstruction in the past can offer any indication, it may be estimated that as high as 60 percent of resources to finance the plan is expected to come from foreign countries. In previous plans, the Soviet Union and China provided the bulk of foreign assistance. Under the current plan, however, it is evident that nonsocialist countries, especially the United States, Japan, and France, and international organizations are counted heavily upon by Vietnam to provide the financing.

Modality of foreign assistance preferred by Vietnam's leaders is, of course, bilateral and on a program basis; that is, the aid donor is expected to supply most of the items related to the project including machinery, tools, materials, and even medicines for workers.

According to American sources, aid from the Soviet Union was in the range of $345-$360 million in 1970 and $315-$450 million in 1971.[21] The economic agreement concluded between the Soviet Union and Vietnam at the end of 1975 envisaged Soviet support for the entire period of the plan, involving the construction of some 40 projects. Most important of all is the hydropower-engineering project on the Black River (Song Lo), which includes a 1.7-million-kilowatt hydroelectric-power plant, a 500,000-kilowatt thermal-power plant, a coal mine with a 2.4-million-ton annual output, and the expansion of the Lao Cai apatite mine to reach 1.5 million tons of production. Assistance will also be provided for oil, gas, and solid-mineral prospecting.[22]

Aid from China, which is estimated to be about $100 million per year in the recent past, may at best be continued at about the same level, given the cooling off of relations between the two countries.

Vietnam can also count on assistance that the former government of the Republic of Vietnam had already negotiated (1974-75) with Japan, France, West Germany, and other Western European countries as well as with the Asian Development Bank. This assistance totals over $100 million.*

The new regime inherited the former Republic of Vietnam's memberships in the International Monetary Fund (IMF) and World Bank in 1975. It received a $36-million loan from the IMF under a "compensatory financing" scheme at the end of 1976, and there was a report that the World Bank was considering a loan of $100-$200 million to finance irrigation projects and railroad modernization.[23]

Vietnam has also asked for aid from the United States,[24] but, as of this writing, the prospects for a large amount of aid are uncertain.

---

*This aid was negoitated during 1974-75 by the author, then minister of economic planning and coordinator of foreign reconstruction aid of the Republic of Vietnam.

In sum, it is possible that Vietnam's planners are thinking in terms of between $1 billion to $1.5 billion for the current plan, averaging $200-$300 million per annum, which is three times higher than the $500-million foreign-aid level projected for the first five-year plan.

CONCLUDING REMARKS

Poised by victory and impressed by Southern natural resources, modern plants and equipment,[25] Vietnam's leaders look to the future with confidence and ambition.

Politically, the country has already been officially unified since the election of the National Assembly in April 1976. Economically, unification is to proceed slowly, but the most important benefit of it from the viewpoint of the North—the flow of food from the Southern delta—has already been resumed. The regime can now concentrate on a new campaign: "all for production, all to build socialism."[26]

The drive began with the intent to implement an indicative five-year plan (1976-80), covering the whole territory of Vietnam. In a broad sense, the plan implies the return to the traditional pattern of specialization as analyzed in the first two chapters: the agricultural South and the industrial North. The only modification to fit the current situation is that the South is to proceed with agricultural processing and with production of some consumer goods (light industry).

The plan also implies the return to the centuries-old pattern of migration: surplus labor in the North to surplus land in the South. In the plan's terminology, the latter is referred to as "regional balance" and "labor force redistribution."

A sharp increase in food-grain production in the South envisaged by the plan is to relieve the North from its eternal food-shortage problem. In addition, Southern agricultural surplus is expected to finance industrialization in the North. The latter is to lead the country politically, ideologically, and to build industry; it will provide the South raw material for manufacturing, fertilizer, and tools for agriculture.

By the end of the current decade, a unified Vietnam is expected to have traversed one-third of the road on its way to becoming almost a utopian modern socialist state, strictly conforming to the Marxist-Leninist perception. The economy will include a modern heavy-industrial base to replace agriculture as its dominant feature; a large class of industrial workers, the proletariat, to replace the peasants as its political foundation; and a strong defense. By 1980, there would also be a clear improvement planned for the Northern people's living standard. The improvement will be made first in basic food and clothing, household needs, public transportation, then, in the long run, in some luxury goods. The plan, as described by Le Duan, envisages the following for the common citizen in the course of its implementation:

> In the next 5 years, we must see that everybody has enough staple food, vegetables, beans, sauce, fish and some of the necessary meat, in addition to sugar, eggs, fruits . . . . We must strive to ensure decent and warm clothes for the people, properly organize transportation, adequately supply all families with common goods such as cooking pots and pans, bowls and plates, tea services, blankets, mosquito-nets, cupboards, beds, tables and chairs, teaching aids, children's toys, etc. We must step by step meet the demands of the people for electric fans, clocks and wrist watehes, radios, television sets, sewing machines. We must attach utmost importance to increasing public welfare establishments.[27]

The target reflects an improvement over past targets envisaged for the three-year plan (1958-60) and the five-year plan (1961-65) in that it includes the provision for electric fans, clocks, radios, and so on, to be realized step by step in the future.[28]

The improved standard, which appears to be high for the Northern population compared to its present level, has long ago been reached in the South; obviously, therefore, there would only be deterioration in the standard of living for the Southern population for the rest of the 1970s.

As with all the previous plans, the current plan appears to be ambitious; that is, the planners want to do everything at the same time—develop rapidly and to the utmost heavy industry, light industry, food grains, industrial crops, forestry, fishing, transport, communications, construction, socialist education, and build a strong national defense.[29] Evidently, even with the highest degree of rigid control and devotion to Communist ideology, the planners cannot escape the awesome choice between "guns and rice" (in Western thought, "guns and butter") on the production-possibility frontier.

In the North, for more than a decade, the choice was overwhelmingly for guns, with very little resources for rice. The result was that, at the dawn of unification, the economic cost of the war becomes more visible than ever before. The Northern economy is in a state of shambles, production stagnant, and the people's standard of living among the lowest, if not actually the lowest, in the developing countries. The past policy of "all for the front, all for victory" has brought about serious misallocation of resources and critical shortages of labor in all fields. The policy of rigid control and of substituting politics for economics in the decision-making process and planning has resulted in confusion, chaos, and inefficiency. Negligible capital investment together with lack of incentive has kept the low level of productivity from rising for decades.

In the South, the nature of war-related problems is different, the most serious ones being the high unemployment level in the urban area, the disequilibrium in production of goods and services, and especially the dependence on foreign resources and markets for industrial production. The availability of foreign exchange in war time also induced fast mechanization of agriculture, which has now become stagnant in the face of a lack of fuel and fertilizer. If stagnation persists in southern agriculture, as it is likely to do, the effects upon the entire economy and eventually upon political stability will be profound indeed.

In addition to the postwar economic conditions, there are other serious problems that will have to be faced. The strength with which North Vietnam fought for victory may turn out to be its weakness after victory has been won: assets in war may become liabilities in peace.

During more than two decades of war, the majority of the male labor force was trained only to fight. It will be difficult now to turn it into a productive labor force. During the war, the entire population was called upon to make tremendous sacrifices. The people now will demand reward. The goal of unification was effectively used to motivate the population to work and to sustain hardship. Unification, now achieved, no longer serves as a useful instrument for mass mobilization. Unification also poses a potential danger to the leaders, for it will expose the Southern prosperity to the Northern population, a fact which until now has been successfully avoided. The serious discrepancy in the standard of living of the two zones, no matter how artificial it seems to be,[30] will tend to weaken North Vietnam's propaganda apparatus and at the same time encourage the Northern population to catch up with the South in consumption to make up for past sacrifices. Finally, the war itself, which was blamed for all the country's economic and social ills in the past, is now over.

Given the country's pressing need for technology and equipment, the opening up of the closed economy for trade with non-Communist countries is inevitable. But import of non-socialist-produced goods and technology also brings with it the labels of capitalism and exposes the fast-changing world to the long-isolated people of the North.

In the South, the sudden disappearance of prosperity, the draining of resources to support the North, the opening up of new economic areas, and forced transfer of labor will certainly cause either active or passive resistance to the new regime. For the first time in history, the population of the Mekong delta is experiencing shortages of food. For the first time they are regimented and subjected to rigid controls and to austerity.

The leaders seem to be fully aware of all these problems. They appear to realize that the time to promise has ended and the time to deliver begins.[31] But whether or not they will actually deliver depends upon their scale of preferences and priorities.

Given the potentials of the unified economy and the inflow of foreign aid, the prospects for the Vietnamese economy hinge not so much on the availability of resources,[32] but rather on how these resources are going to be employed; that is, what mixture of economic well-being of the population on the one hand and socialist building and defense on the other the planners are going to take.

Under the 1976-80 plan at least, the present leaders seem to be caught in the ideological trap and appear to have opted for maintaining, in essence, the old line of the party. But it becomes uncertain now more than ever before whether the status quo will necessarily endure. In addition to the economic forces, there are other socio-politico-economic factors that accompany unification. The rising power of the armed forces, the growing influence of the younger generation, which did not participate in the 1945 August revolution, the Southern active and passive resistance to the new regime, the reactivation of problems of regional and religious divisiveness can all be expected to generate overwhelming internal pressure to influence the future course of development of the unified Vietnamese economy.

As the country faces an uncertain future, one thing seems certain: the close relationship between the economic factor and the rising power of the Communist party. It was the food crisis that brought the party to power in 1945; it was food pressure in the North combined with a promise to discharge this pressure with the resources of the South that helped the party to mobilize the population for total victory in 1975. Toward the future, it seems relevant to quote a remark by the late Bernard Fall, a well known authority on North Vietnam:

> On its the Communist party's eventual success or failure in providing people under its control with an adequate living may well depend in great part the survival of the whole regime.[33]

NOTES

1. Douglas Pike, "Vietnam: Democratic Republic of Vietnam," in Yearbook on International Communist Affairs, 1976 (Stanford: Hoover Institution, Stanford University, 1976), p. 395.

2. Che Viet Tan, "Developing the Combined Strength of the Unified Economy," Hoc Tap, no. 12 (1975), translated by U. S. Department of Commerce, Joint Publications Service, Translations on North Vietnam (hereafter cited as TNVN), no. 1770, February 17, 1976, p. 18.

3. Ibid., pp. 18-19. See also Le Duan, "Toan Dan Doan Ket Xay Dung to Quoc Vietnam Thong Nhat Xa Hai Chu Nghia" (The entire people unified to build the reunified socialist Republic of Vietnam) (Speech delivered to the National Assembly, April 25, 1976), Hoc Tap, no. 7, 1976, p. 30.

4. Le Duan, "Report to the Fourth Congress, Vietnam Workers Party," Radio Hanoi broadcast, December 16, 1976, published in U. S. Foreign Broadcast Information Service, APA 4, no. 247, supp. 48, December 22, 1976, p. 37.

5. Ibid., pp. 32-33.

6. It seems relevant to note here that the 1976 state plan for North Vietnam assigned a target growth rate of 15 percent for the year, of which paddy was to increase by 13 percent (to 5.5 million tons), and subsidiary crops, 26 percent. The seemingly high growth rate projected for 1976 is explained by the poor performance in the agricultural sector during 1975. Compared to 1974 (a good year), the 1976 food-grain output target would have been substantially smaller (2 percent). Le Thanh Nghi, "Tinh Hinh Kinh Te Mien Bac Hien Nay Va Nhiem Va Ke Hoach Nha Nuoc Nam 1976" (Economic situation in the North and the tasks and plan for 1976) (report given at the Second Session of the National Assembly, Fifth Legislature, December 1975), Hoc Tap, no. 1, 1976, pp. 16-17.

7. Land under forest cultivation is estimated at 15 million hectares. (Hoc Tap, no. 12, 1975, TNVN, no. 1770, op. cit., p. 17.)

8. For a treatment on this subject, see Doan Dinh Hoe, "Phân Vung Kinh Tế ở Vietnam" (Economic zoning in Vietnam) Hoc Tap, no. 4, 1976, pp. 55-64.

9. See Le Duan and Pham Van Dong, Vers Une Agriculture de Grande Próduction Socialiste, (Hanoi: Editions En Langues Etrangeres, 1975).

10. Viet Tan, op. cit., p. 21; Duan, "Report to the Fourth Congress, Vietnam Workers Party," op. cit., p. 46. For a background study on new economic zones in North Vietnam, see Le Hong Tam, "Ve Van De Xay Dung Cac Vung Kinh Te Moi" (On the question

of developing new economic zones), Hoc Tap, no. 6, 1972, pp. 48-59.

11. Viet Tan, op. cit., p. 23. For overall guidelines in agricultural development, see Duan, "Report to the Fourth Congress," pp. 33-34.

12. Viet Tan, op. cit., p. 20.

13. Ibid.

14. Duan, "Report to the Fourth Congress," p. 37.

15. Ibid.

16. Viet Tan, op. cit., p. 19.

17. Le Duan, "Report to the Fourth Congress," op. cit., p. 38.

18. Ibid., p. 41.

19. Ibid., p. 45. On the need to broaden the foreign trade base, a North Vietnamese was more candid:

> Building an independent, autonomous economy does not mean that we close the door and become self-sufficient. Under the socialist system, the more independent and autonomous the economy becomes, the better able we are to widen our economic relations with foreign countries; conversely, as our economic relations with foreign countries widen, the more we can consolidate our independent, autonomous economy; these two aspects supplement each other and create the conditions for each other's development. By widening our economic relations with foreign countries, we can supplement the weaknesses and develop upon the strengths of the country's economy . . . . The thing our country lacks the most in its advance from small-scale production to large-scale socialist production is large-scale industry, the material and technical base of socialism. Therefore, it is necessary to endeavor to widen our economic relations with foreign countries in order to rapidly build the material and technical base of socialism.

Nguyen Can, "Researching the Revolutionary Situation and Task in the New Stage," Hoc Tap, no. 12, 1975.

20. Duan, "Report to the Fourth Congress," p. 44.

21. See Business International Asia/Pacific Ltd., Risks and Rewards in Vietnam's Markets (Hong Kong: Business International Asia/Pacific Ltd.), p. 177.

22. It has also been reported that the Soviet Union will provide a glass factory at Cam Ranh Bay. See Ellen J. Hammer, "Communist But Nonaligned," Problems of Communism (U.S. Information Agency) 25, May-June 1976, p. 8.

23. Washington Post, March 3, 1977, p. A-1; Wall Street Journal, March 11, 1977, pp. 1, 24.

24. Washington Post, January 7, 1977, p. A-1, March 18, 1977, p. A-24; Wall Street Journal, March 11, 1977, pp. 1, 24.

25. The value of the machines, plants, and equipment in the South, which existed at the time of the Communist takeover, was $12 billion; that of military hardware, $6 billion (Hammer, op. cit., p. 400).

26. Le Duan, "Toan Dan Doan Ket Xay Dung To Quoc Vietnam Thong Nhat Xa Hoi Chu Nghia (The entire population unified to build the reunified socialist republic of Vietnam) Hoc Tap, no. 7, 1976, p. 46.

27. Duan, "Report to the Fourth Congress," p. 48.

28. Target standards in previous plans included improvement in basic food, clothing, housing, education, and health. See Plan Triennal (Hanoi: Editions en Langues Etrangeres, 1959), pp. 29, 38, 40; Vo Nhan Tri, Croissance Economique de la Republique Democratique du Vietnam (1945-65) (Hanoi: Editions en Langues Etrangeres, 1967), p. 374.

29. Duan, "Report to the Fourth Congress," p. 32. The question of coordination of economic development and national defense is explained this way:

> The economy creates and supplies material forces for war. The economic system has a very great influence upon the fighting morale and the ability to develop the political and spiritual strength of the army and the people in a war. The economy is very closely related to the defense of the nation. As a result of thoroughly implementing the guidelines of "producing and fighting," "a plow in one hand, a rifle in the other," and "a hammer in one hand, a rifle in the other," our people defeated two powerful enemies, the French imperialists and the U. S. imperialists. Therefore, in order to protect the fatherland so that it is forever independent and free, we must fully comply with the guideline of closely coordinating the economy with the defense of the nation and make every effort to increase the economic and national defense potentials of our country in order to insure the completion of economic tasks and meet the requirements of the task of consolidating and strengthening the national defense system.

Nguyen Can, "Researching the Revolutionary Situation and Task in the New Stage," Hoc Tap, no. 12, 1975.

30. Evidently conscious of the problem of great discrepancy between the living standards of the North and the South, Duan warned that those who have had an artifically high standard of living due to the occupation of the American imperialist in the past must now return to reality (Duan, "Toan Dan Doan Ket Xay Dung To Quoc Vietnam Thong Nhat Xa Hoi Chu Nghia," pp. 31-32.)

31. In his will, Ho Chi Minh decreed: "Our mountains will always be, our rivers will always be, our people will always be. The American invaders defeated, we will build our land ten times more beautiful" (Ho Chi Minh, *Selected Writings*, op. cit., p. 361.) The quotation is now repeated often by the Communist leaders. See Duan, "Toan Dan Doan Ket Xay Dung To Quoc Vietnam Thong Nhat Xa Hoi Chu Nghia op. cit., p. 46; Truong Chinh, "Cuoc Tong Tuyen Cu 25-4-1976 Danh Dau Buoc Thang Loi Quy et Dinh Cua Nhan Dan Ta Tren Con Duong Thong Nhat Nuoc Nha ve Mat Nha Nuoc" (The April 25, 1976 election marked the decisive victory of our people on the way to unify the country at the state level), *Hoc Tap*, no. 7, 1976, p. 23; Pham Hung, "Mien Nam Nguyen Dem Het Tinh Than va Nghi luc Cung Dong Bao Ca Nuoc Giu Vung Hoa Binh, Cung Co Doc Lap, Hoan Thanh Thong Nhat" (The South pledges to mobilize total spirit and energy to join the nation in maintaining peace, consolidating independence, and reunifying the country), *Hoc Tap*, no. 7, 1976, p. 49.

32. One report suggests that Vietnam may have received promises of more than $6 billion in economic aid over the past two years (Washington *Post*, January 21, 1977, p. A-3).

33. Bernard B. Fall, *The Viet-Minh Regime* (New York: Institute of Pacific Relations, 1956), p. 98.

ABOUT THE AUTHOR

G. NGUYEN TIEN HUNG is currently Associate Professor of Economics at Howard University in Washington, D. C. Formerly an economist for the International Monetary Fund in Washington, D. C., professor Hung has taught at North Carolina Wesleyan College and Trinity College. He received his B. A. from Saigon University and his M. A. and Ph. D. in economics from the University of Virginia. He has followed the economic problems of North Vietnam for many years and written extensively on the subject. After the ceasefire of January 1973, he returned to South Vietnam and became Minister of Economic Development Planning in charge of postwar reconstruction. His other publications include works on development planning and financing in developing countries.